Shobal Vail Clevenger

Comparative Physiology and Psychology

a discussion of the evolution and relations of the mind and body of man and

animals

Shobal Vail Clevenger

Comparative Physiology and Psychology
a discussion of the evolution and relations of the mind and body of man and animals

ISBN/EAN: 9783337239381

Printed in Europe, USA, Canada, Australia, Japan

Cover: Foto ©berggeist007 / pixelio.de

More available books at **www.hansebooks.com**

COMPARATIVE

Physiology and Psychology.

A DISCUSSION OF THE EVOLUTION AND RELATIONS OF THE MIND
AND BODY OF MAN AND ANIMALS.

By S. V. CLEVENGER, M. D.,

SPECIAL PATHOLOGIST COUNTY INSANE ASYLUM, CHICAGO; MEMBER OF THE AMERICAN
NEUROLOGICAL ASSOCIATION, AMERICAN MICROSCOPICAL SOCIETY, AMERICAN ELEC-
TRICAL SOCIETY, AMERICAN ASSOCIATION FOR THE ADVANCEMENT OF
SCIENCE, CHICAGO MEDICAL SOCIETY; EX-METEOROLOGIST U. S. SIGNAL
SERVICE; COLLABORATOR OF THE AMERICAN JOURNAL OF
NEUROLOGY AND PSYCHIATRY, AMERICAN JOURNAL
OF NERVOUS AND MENTAL DISEASE, AND
AMERICAN NATURALIST;
ETC., ETC.

CHICAGO:
JANSEN, McCLURG, & COMPANY.
1885.

Cowdrey, Clark & Co., Printers,
Chicago.

PREFACE.

Some of the original ideas contained in this book have appeared in scientific and medical publications, such as the *American Naturalist* and *Journal of Neurology and Psychiatry*, during the past five years, as the author presented the papers containing them to biological, microscopical, medical and general scientific societies. In their condensed form, herein, the separate theses are revised and amended in conformity with more recent psychological and anatomical research.

About eleven years ago the plan for these investigations was formed and has since been consistently pursued through most discouraging difficulties. All the notes accumulated could not be compressed into a volume of this size, and the author was compelled to content himself with including enough of the mental operations of man to fairly illustrate the comparative method, which will again be applied especially to the mechanism of the mind of man in a forthcoming work to be entitled "Psychology."

Personally made studies of savages, infants, and all classes of men living in so-called civilized communities, with his published and unpublished clinical and pathological reports of cases of insanity enable the author to advantageously review the literature of psychology and psychiatry.

His intention is to elaborate, as far as possible, a practical mental science which will reconcile the observations of anatomists, psychologists and pathologists with direct reference to the more intelligent treatment of insanity.

The author has been greatly assisted in his labors by Dr. E. C. Spitzka, Professor of Neuro-Anatomy and Pathology at the New York Post Graduate Medical School, and Dr. James G. Kiernan, Medical Superintendent of the County Insane Asylum, Chicago: two men (in the exalted sense of the word), whose efforts, to advance the science of psychiatry and render more humane the treatment of the insane, have been a prolonged but successful struggle against political and general greed, ignorance and malevolence.

Faraday, Huxley and Tyndall in chemistry, biology and physics, with the host of workers in nerve phenomena have afforded the materials for the author's work.

Charles Darwin and Herbert Spencer have taught him how to make use of them.

TABLE OF CONTENTS.

Comparative Physiology and Psychology.

CHAPTER I.

INTRODUCTION.

Insanity will be better understood and its treatment will become more scientific in proportion to the development of psychology, based upon comparative microscopic anatomy and a physiology into which molecular physics shall enter more in the future.

Heretofore the mental workings have been discussed chiefly by a class called metaphysicians, many of whom were astute observers; but in the main their system was so insufficient, so one-sided, and their deductions often so absurd as to discourage honest investigators and cause such a thing as a science of the human mind to be looked upon as chimerical, and even though possible to achieve, as of doubtful use.

The study of the mind has thus fallen into disrepute among many scientists, and as has been the case with all branches of knowledge, it has been travestied by such imposters and igno-ramuses as phrenologists, spiritualists, mind-readers, magneti-zers, pseudo-psychologists, and has been honestly assailed by a sprinkling of bewildered metaphysicians, with rarely, here and there, physiologists, such as Carpenter, Maudsley, etc.

It is not intended here to review the matter historically. In-cidentally it will appear how vast has been the influence of

Herbert Spencer and Charles Darwin in the evolution of a sensible psychology, but the work done by Huxley, Wundt, Ribot, Meynert, Spitzka and a few other special workers in biological fields should be accorded full recognition.

The method by which I propose to examine the mind is an extension of Herbert Spencer's principles. Analytical and synthetical applications of many fields of research, mainly physiological. Comparative embryology and physiological chemistry are rapidly clearing the way for safer inductions, though alone they would be insufficient for our purpose. Catholicity and the grasp of a multitude of scientific facts, apparently having but little to do with the subject, will give succeeding generations safer and still safer bases for the coming comparative psychology, which will rescue both mental study and general medicine from empiricists and impostors.

It may be startling to many of my readers to learn that the drift of physiological inquiry has been steadily toward the recognition of sensation and memory and all the vast subsidiary mental processes, correctly and incorrectly called emotions, feelings, cognitions, etc., as modes of chemical energy.

Chemical union and interchange of atoms is conceded by all, but the abjectly ignorant, in other life phenomena, as in assimilation, blood, bone, cartilage and muscle-building; but the nervous system, through effects of olden superstitious training, has been regarded as in some way exempt from the full operation of natural laws, and the brain is looked upon by many sincere and otherwise well-informed persons, to-day, as a meeting place of material and inscrutable spiritual laws.

I wish to be understood by all classes of thinkers, for a fair beginning will acquit me of ambiguity.

That mind is chemical affinity, or allied to it as a property of matter, seems a terribly blasphemous assertion to many of the old superstitious way of thinking. To such who cannot rid themselves of their fetiches and whose bias prevents their intelligent examination of any physiological study, I have nothing to say. If they are content with ignorance they must make the most of it; but with all who are not afraid to follow wherever facts will lead them, whatever their inclination to believe may be, we may meet upon the common ground of acknowledging that there is an UNKNOWN. I will not grant an *Unknowable,* for to set limits upon what may be disclosed to us in the future argues a knowledge of where science will always fail, which no one possesses. That is, we cannot say that we will never know certain things, for each century moves the preceding limits of the known farther into what has been before the realm of the unknown. At present it would be arrogant indeed to say that everything was known, and hence I start with the full admission of *ignoramus* as to ultimates, but by what right can we cry *ignorabimus?* Knowledge being relative, who shall fix the boundaries of the ultimate?

Dualists and the ecclesiastically biased, who are anxious to throw the responsibility of the universal workings upon a capricious influence of which we know nothing, should be content with this acknowledgment of an unknown beyond. With this unknown we shall have nothing whatever to do, and we will proceed at once to consider *what is known* and to argue therefrom.

If the monistic philosopher, who recognizes the guidance of one universal set of laws in nature, be told that he is assuming when he degrades mind to the level of chemical affinity, let him reply that in the first place it is assumption indeed to regard chemical affinity as a degraded thing. Rather let mind and chemical affinity be alike considered as *in their essence* unknown; let them be honestly regarded as alike in the matter of present understandability, and admitting that we do assume in claiming this relationship of mind and chemical affinity, we claim that the assumption will be fully justified by the deductions.

A postulate is absolutely necessary to an argument, and you cannot move in any direction without one. You cannot step out of your house without having premised that the street exists, nor can you retrace your steps without the supposition that your house will be found where you left it.

The assumption of mind having evolved from chemical affinities and being related to them affords us somewhat, and a somewhere, upon which to advance; and, as we advance, the consistencies are so beautiful and the inferences so unerring it seems inexpressibly stupid that we should have been so long in the dark.

The entire fabric will be a triumph of monism, for if we set out on any other assumption, such as the dualistic affords, than that mind is a product of chemical energy and other natural forces there is an end to inquiry.

The baleful influence of teleology hangs over the average physiologist as over the superstitious laity, and debars him from seeing things as they really are. The inability to conceive of consciousness as a product of the motions of matter is on a level with the inscrutability of the nature of ultimate force and atoms. In dealing with the workings of the mental mechanism it is not necessary to define or attempt to explain consciousness, any more than the practical electrician or chemist or optician finds it necessary to define or speculate upon the ultimate nature of the vibratory terms in which they deal. As the physicist increases his knowledge of how matter and motion act and react upon each other, he is willing the metaphysicians should quarrel over the unknowable, the lunar politics. With the dawn of comparative psychology the truth began to appear, theories became subordinated to facts and not facts to theories.

Not only are the laws which bind the social organism similar to and derived from those which govern the units of which it is composed, but the protoplasmic units are governed by the same processes down to chemical affinities.

H. C. Sorby* estimated the number of molecules in 1-1000 inch sphere of albuminous substances to be:

Albumen................................. 10,000,000,000,000
Water........................... 520,000,000,000,000
Water in molecular combination.....530,000,000,000,000

and claims that we are as far from seeing the ultimate constitution of organic matter with our highest and best powers as the naked eye is from seeing the smallest objects which they now reveal to us, and there seems no hope that we may ever see them, for light is too coarse.

This is a limit to our sense appreciation of the subject. Reason enables us to make safe guesses beyond the senses, but having them for a guide and acknowledging that science does not require final but effective causes.

* Presidential address, Royal Micros. Soc., Feb. 2, 1876.

CHAPTER II.

PRIMITIVE LIFE AND MIND.

The "selection" of food which is suited to the amœba becomes "selection" or mere chemical attraction, depending upon how you look at it. For instance, let the assimilable pabulum consist of molecules for which the protoplasm has affinities or attractions. The amœba will not only be drawn to it locomotorially, but will fuse about it. As it is drawn into and becomes part of its tissue, there is undeniable chemical union. The inert resulting matter is left behind or excreted in the movements. The inert matter not only does not attract the animal, but even in passing to or over it the assimilative motions are not provoked. There could be an endless wrangle over the nature of this act of the protozoon for it involves the most weighty considerations in all life. There is much to be said on all sides, but the moment the acknowledgment is made that chemical affinity and other physical influence is not the so-called will power of the amœba, that moment there is an end to investigation. Admit that these natural causes exert entire control of the protozoon, and forthwith the postulate proves its correctness in exact proportion to the correctness of the logical methods used in reasoning therefrom.

Much of the perplexity into which the student has been thrown by regarding these movements, has arisen from want of consideration of the composition of resultants of attraction from many points in the medium or environment due to light, heat, eddies, vortices, disseminated invisible attracting points, the assimilative process itself changing the conditions of attraction. In plethora, as might be expected, movements cease, owing to combinations being satisfied for the time. Then, too, the simple nature of protoplasm has been by no means proven. It is being regarded as not only complex through atomic union, but as holding in its molecular construction secrets which the chemist may some day find operative in the inorganic affinities. There may be, as has been surmised, many kinds of protoplasm, and the ultimate basis-substance may be beyond. The ova of the different animals seem to be protoplasm plus other things, differing from each other in quantities and composition. We know that certain animals add to their bodies chemical substances which form tissues, and that other animals do not, showing a variability of selective affinity. The psychologist who attempts to explain consciousness on the basis of molecular reaction is no more at a loss than the chemist who accepts such words as catalysis and isomerism as representing acts of the atoms.

Starting out, then, with the fair understanding that the amœba moves by virtue of the operation of physical causes, and that speculations upon the origin of matter and force are foreign to the subject, we will see to what the assumption, if you choose to call it one, will lead.

I invite earnest attention to the proposition I make here as corollary from the apparent volition of the amœba being molecular attraction. *Locomotion and prehension of the amœba are due mainly to extrinsic forces operating immediately upon the organism. Whereas these phenomena in man and the higher intermediate metazoa are due immediately to intrinsic forces, as a rule, preponderating over the extrinsic, but nevertheless the extrinsic remain the remote causes of motion in all animals.* In this there is a view of the evolution of volition from the so-called involuntary, its growth from the chemical affinities.

The belief is current among biologists that if we reverse the conditions under which all life exists, all life would perish; if the reversal were slowly effected, most would perish and but few survive; if inappreciably slowly, it is highly probable that the number of surviving forms would be very large. The survival of any animal is evidence of its consonance with its surroundings, and the environment not only modifies and acts upon the animal to develop or destroy it, but also from our chemical standpoint originates it. Without dipping into biogenesis, spontaneous generation somewhere, somehow, is consistent with the ground-work of our essay, but we will avoid its consideration.

Spallanzani, Dugé, Doyère and others have demonstrated that infusoria and certain low worms, the rotatoria, tardigrada and some crustacea are capable of dessication and revival. The suspension of the major evidence of life function by animals under changed conditions, whether this be absolute dessication or not, the development of seeds and ova after indefinite quiescence, point toward, if they do not fully attest, the merging of the inorganic into the organic and the addition of the faculty known as life through the restoration of the medium, water, which affords the means for the molecular motions which go to make up all there is in life.

So the restoration of frozen fish, and as Semper cites, "Amphibia, Mollusca and other forms have lived years without food." He "kept species of landsnails for years wrapped in paper and quite dry in wooden boxes and thus wholly without food, and many of them are at this day alive and active." His explanation is: "The amount of nourishment required daily by any animal must naturally be equivalent to the organic matter which is daily used up in the various organs to keep up the vital processes. The more active an animal is the more food will it require. But the vital processes of animals as low in the scale as amphibia or univalves are extremely feeble; their respiration even under the agitating influence of propagation fails to raise temperature appreciably. In such the vital process may be reduced to a minimum without loss of life."

The whole matter is one of degree, for warm-blooded animals live but a little while unfed. Hibernates are comparable in condition to "cold-blooded," while this division sustain arrest of nutrition longer, and finally in the lowest forms the approach toward almost indefinite suspension leads us to think that there is a point where life and mere chemical conditions are identical: the repeated withdrawal of that which renders life evident entailing no permanent inconvenience. Such embryos as are capable of living in a medium such as strong alcohol several days, point to the mechanical nature of certain low stages of life and the diminished liability to destruction of initial forms through heterogeneity of environment. The internal tissues of man, with their great range of chemical natures of fluids in which the cells thrive, instance this. With differentiation and higher organization comes the increased necessity for stability of environment paralleled by the ability of low forms to reproduce lost members, not evident in developed life.

We may regard the amœba in many ways as having undergone development above some lower forms, but pending the settlement of the bathybius question and with a mere glance at protista and pre-amœbic life, this organism affords us a convenient starting point for inquiry. While physiologists agree in its possession of the fundamental activities of life in simplest modes of manifestation, they usually content themselves with a mention of this fact and proceed to examine complex differentiated tissues, as though the amœba merited no further attention.   From my way of looking at it, the amœba, containing the solution of so much, deserves very deep consideration, which being accorded it the apparently simple becomes intricately complex in that it explains so much.   First, the environment of the amœba: stagnant water, mud or damp earth, or from the infusion of any animal substance in water and allowing it to evaporate while exposed to direct sunlight.*   It absorbs oxygen and gives out CO_2 at 45° C., and strong shocks of electricity kill it.   Moderate shocks of electricity cause it to assume the globular form.   Crushing kills it, and then even the nucleus disappears.   Freezing point arrests its motions.   In its surroundings there are, besides its food, air, water, mineral matter, sunlight, heat and cold, and mechanical vibrations.

At 35° it heat stiffens, at once proving the development of the amœba for its medium and that of the white blood corpuscle, which is more sluggish, for a different one, the temperature of the blood currents of the different animals.   This may be regarded as an acquired adjustment.

Its molecules are subject to the laws of gravitation; light attracts it; heat increases, within limits, its activity; vibrations, such as eddies of its medium, move it; electricity stuns it; its intimate structure assimilates chemically the substances for which its molecules have affinities, and being nonresponsive to those for which it has not, consigns them to the exterior.

Now, if all these forces act upon and in the amœba, what is to prevent external forces from pulling or pushing out its pseudopodia and with the cohesion of its mass flowing its granules into the pseudopodium most attracted, and thus its being drawn bodily in the line of the resultant of all its

---

* Practical Biology, Huxley & Martin.

external and internal forces. The multiplicity of the components of the resultants are evident upon watching it.

However highly differentiated the desires of man may be, and however he may fail to recognize the attraction of his own cells for pabulum, as soon as the food is placed within reach of the enteric cells *their* affinities are not masked. If a complex organic protoplasm has the capacity of chemical conversion and union with oxygen and other molecules, and at the same time the union of oxygen and hydrogen under proper circumstances is through such conditions favoring the mutual attraction at a distance, we cannot avoid the idea that a similar effect is produced upon the bioplasm, and that affinity for its food is a chemical energy, which is *one* of the forces forming with other modes of motion, or attraction, a resultant; each of these attracting inversely as the square of its distance and directly as its mass. Because the protozoon does not go straight towards its food it is thought not to be attracted by it; but when in contact the pseudopodia envelop it, then it is said to be a will effort. When in contact, then the assimilation is possible, and chemical energy asserts itself as a larger component of the general forces which make resultant motions; when at a distance, the food becomes *one among many* influences upon its movements.

Prehension, which is here evidently locomotory, is for the obtaining of food, and is caused by a number of natural extrinsic forces or attractions combined with a lesser number of intrinsic forces or attractions.

Chemical affinity is the prime cause of assimilation. Locomotion is evidently here only a form of the latter, due to the former as a direct cause, but accidentally aided; often interfered with, by other similar natural forces, inasmuch as the amœba may be drawn away from its food, unless it be near enough or there be a compensatingly large enough amount to draw it against opposing attractions.

Throughout animal life, to the highest, with the development of food-procuring faculties this rule still holds good. The more the faculties increase, the more direct is the food acquisition, and the less do generally coöperative, but in this regard interfering processes influence food-procuring. Atavism is prominent in doing that which drives from a base of supplies or want of foresight in improvidence.

Prehension is an accessory to locomotion and both are assimilative acts, or acts which have for their end the assimilative.

This is also evident in full development, for every act or movement of the whole body is of a prehensile nature,—leg movements take hold of the ground through gravitation to carry the body in search of food; hands and arms being prehensile direct; jaws are prehensile in their food grasp; ribs are prehensile in their assisting oxygen introduction, that gas being a food. In snakes the ribs are locomotory prehensile.

We thus have all the physical forces, including gravitation and chemical energy, acting upon the low organism to cause *all* its motions. Just as the heat of the sun overcomes the earth's gravity, and lifts billions of tons of water from the ocean to allow it to fall again in obedience to terrestrial attractive force, so may the "vitality" of an animal or plant *apparently* working against physical laws, lift the child from the embryo, the tree from the seed. But eventually the cycle is complete and the primitive elements are separated in "death," to re-enter at once upon other changes. The natural forces are masked in life phenomena, as the law of gravitation, though the direct agent, is not recognized in the upward rush of the fountain.

The amœba assimilates organic matter and breathes as it uses up oxygen and exhales carbonic acid. To complete the ob-

jective study of the amœba we observe that it *grows* as a consequence of its eating, and that, owing to its growth and the
operation of the attraction of gravitation, a force too often
neglected in consideration by physiologists, fission or reproduction occurs, as the cohesive attraction of its molecules cannot pass beyond a certain limit, and the extra weight is *gravi-
tated*, excreted off, a process still evident in all animal reproduction and through excretory channels also. We see that
the sexual act is identical in this so-called neutral form. This
is more apparent in other protozoa as a differentiation. I ap-
pend my article on this subject from *Science* (N. Y.), June 1,
1881.

"A paper on Researches into the Life History of the Monads,
by W. H. Dallinger, F. R. M. S., and J. Drysdale, M. D., was read
before the Royal Microscopical Society, Dec. 3, 1873, wherein
fission of the Monad was described as being preceded by the
absorption of one form by another. One Monad would fix on
the sarcode of another and the substance of the lesser or under one would pass into the upper one. In about two hours
the merest trace of the lower one was left, and in four hours
fission and multiplication of the larger Monad began. A full
description of this interesting phenomenon may be found in
the *Monthly Microscopical Journal* (London), for October, 1877.

" Professor Leidy has asserted that the amœba is a cannibal,
whereupon Mr. Michels, in the *American Journal of Microscopy*,
July, 1877, calls attention to Dallinger and Drysdale's contri-
bution, and draws therefrom the inference that each cannibalistic act of the amœba is a reproductive, or copulative one, if
the term is admissable. The editor (Dr. Henry Lawson), of
the English journal agrees with Michels.

"Among the numerous speculations upon the origin of the
sexual appetite, such as Maudsley's altruistic conclusion, which
always seemed to me to be far-fetched, I have encountered
none that referred its derivation to *hunger*. At first glance such
a suggestion seems ludicrous enough, but a little considera-

tion will show that in thus fusing two desires we have still to get at the meaning and derivation of the primary one—desire for food. The cannibalistic amœba may, as Dallinger's Monad certainly does, impregnate itself by eating one of its own kind, and we have innumerable instances among algæ and protozoa of this sexual fusion appearing very much like ingestion. Crabs have been seen to confuse the two desires by actually eating portions of each other while copulating, and in a recent number of the *Scientific American*, a Texan details the *Mantis religiosa* female eating off the head of the male mantis during conjugation. Some of the female *Arachnida* find it necessary to finish the marital repast by devouring the male, who tries to scamper away from his fate. The bitings and even the embrace of the higher animals appears to have reference to this derivation. It is a physiological fact that association often transfers an instinct in an apparently outrageous manner. With quadrupeds it is most clearly olfaction that is most related to sexual desire and its reflexes, but not so in man. Ferrier diligently searches the region of the temporal lobe near its connection with the olfactory nerve for the seat of sexuality, but with the diminished importance of the smelling sense in man the faculty of sight has grown to vicariate olfaction : certainly the ' lust of the eyes ' is greater than that of other special sense organs among Bimana."

" In all animal life multiplication proceeds from growth, and until a certain stage of growth, puberty, is reached, reproduction does not occur. The complementary nature of growth and reproduction is observable in the large size attained by some animals after castration. Could we stop the division of an amœba, a comparable increase in size would be effected. The grotesqueness of these views is due to their novelty, not to their being unjustifiable.

" While it must thus seem apparent that a primeval origin for both ingestive and sexual desire existed, and that each is a true hunger, the one being repressible and in higher animal life being subjected to more control than the other, the question then presents itself : What is hunger ? It requires but little reflection to convince us of its potency in determining the destiny of nations and individuals and what a stimulus it is in animated creation. It seems likely that it has its origin in the atomic affinities of inanimate nature, a view monistic enough to please Haeckel and Tyndall."

Dr. Spitzka, in commenting on the foregoing in the same journal, June 25, 1881, says:

" There are some observations made by alienists which strongly tend to confirm Dr. Clevenger's theory. It is well known that under pathological circumstances relations, obliterated in higher development and absent in health, return and simulate conditions found in lower, and even in primitive forms.

"An instance of this is the pica or morbid appetite of pregnant women and hysterical girls for chalk, slate pencils and other articles of an earthy nature. To some extent this has been claimed to constitute a sort of reversion to the oviparous ancestry, which, like the birds of our day, sought the calcareous material required for the shell structure in their food. (?) There are forms of mental perversion properly classed under the head of the degenerative mental states with which a close relation between the hunger appetite and sexual appetite become manifest.

" Under the heading ' *Wollust, Mordlust, Anthropophagie*' Krafft-Ebing describes a form of sexual perversion where the sufferer fails to find gratification unless he or she can bite, eat, murder or mutilate the mate. He refers to the old Hindoo myth Çiva and Dúrgá as showing that such observations in the sexual sphere were not unknown to the ancient races. He gives an instance where, after the act, the ravisher butchered his victim and would have eaten a piece of the viscera ; another where the criminal drank the blood and ate the heart ; still another, where certain parts of the body were cooked and eaten." *

* Ueber gewisse Anomalien des Geschlechtstriebes, Von Krafft-Ebing, Archiv fr Psychiatrie, VII.

Nature (London), commenting on my article, quotes : " *Mulieres in coitu nonnunquam cervicem maris mordunt.*"

The locomotory, which exhibits all the prehensile acts undifferentiated, is a product of a number of natural forces and, so far as we can speak of atoms having objects, the object of locomotion is in food procuration.

Prehension, locomotion, assimilation, growth, excretion, reproduction are so combined as to appear inseparable. All are molecular motions, integrating to form mass motions, and the latter to facilitate the first. Keeping this in sight as a biological fact, it will simplify subsequent inquiry.

Adjustment and readjustment of the animal perpetually occurs. The reaction of the protozoon upon its environment is possible only through the intimate structure of the animal having been modified by the environment. This consists in molecular changes, ending in mass changes.

This tendency is exhibited in the frequent appearance of a part unable to throw out pseudopodia, which gravitates to the rear, and thus becomes the hinder part. This occurs in the proteus animalculæ temporarily and in other amœba forms as a permanent differentiation. The ectosarc (Fig. 1), or

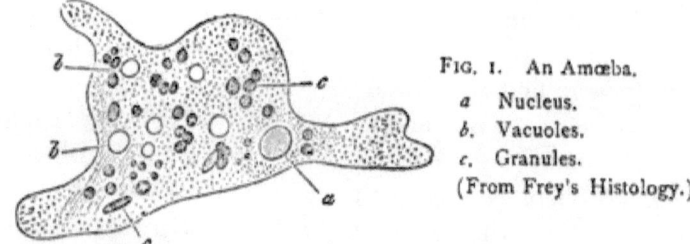

FIG. 1. An Amœba.
a   Nucleus.
b. Vacuoles.
c. Granules.
(From Frey's Histology.)

outer envelope, is composed of denser but mobile materials, due to separation of granular and molecular matter by natural causes.

The vacuoles, or little transparent points, are, with reason, assumed to be watery or gaseous spots, filtered from the assimilative process. Their constant appearance and disappearance are doubtless chemical and mechanical. The CO_2 and water holding in solution or suspension fine excretory material will find its way out through diffusion and the elasticity of the sarcode, with other large particles gravitating out through any temporary channel. This process is apparent, though better provided for, in cloacal animals, whose watery, gaseous and solid excreta are poured forth from fixed, often the same, orifices. The gastræa stage is the condition in full. If, with ingestion of food and oxygen, the animal increase its bulk faster than the ectosarc can accommodate itself to the change, extraneous matter, such as carbonic acid and water, *must* be propelled away from the protoplasm, for which it has no affinity; and under the operation of incessantly recurring similar causes it is not surprising that this rythmical diastole and systole should often become quite regular.

We thus have inspiration of oxygen as an assimilative act in its affinity for the protoplasmic molecules, together with the other accretive atomic motions and elasticity of the ectosarc, instituting rythmic contractions to expel inert products. If this be admitted, then the inspiratory oxygenation of every enteric and arterial cell from the food and blood is the direct cause of vermicular motion and pulsation.

The motions of the amœba are assimilatory, prehensile, locomotory, accretory, inspiratory, expiratory, excretory, reproductory.

Turning now from the objective method, let us examine this primitive form subjectively. The objections to an application of the latter process to amœbic movements are equally valid against all other animals, even man. We know nothing of the workings of consciousness in others except by comparing like effects and inferring from them similar causes. We have the

2

various molecular and molar workings of the amœba as a guide in determining what it feels, likes and dislikes.

Descartes conclusion, "*Cogito ergo sum,*" Huxley regards as *non sequitur.* I would merely postulate both ends of the sentence as being, for physiological study, unassailable: *Sum et cogito,* and let the metaphysicians wrangle over the rest. The amœba's functions are simple but nevertheless the same as our own. Forthwith we must assign it a desire for food, which desire is the chemical affinity of atoms; then the amœba hungers.

Prof. E. D. Cope* assigns it as "the primitive desire and a form of pain. This was followed by gratification, a pleasure, the memory of which constituted a motive for a more evidently designed act, viz.: pursuit."

Dividing these primitive desires arising from (or with, if you wish,) the atomic affinities into those which subserve and those which oppose assimilative processes, we have the origin of pain and pleasure, under which two heads all conscious workings may be classed. Pain increases with the quantity of atoms unsatisfied. As long as there are protoplasmic molecules with affinities, the number of them wanting food increases the desire. (Attraction directly as the mass). Of course, as soon as destructive starvation breaks down the molecule the desire ceases. This is evident in the final loss of desire for food in extreme deprivation in man.

All unsatisfied desire is painful, as:

Hunger in the absence of food;

Desire to move about while disabled from so doing;

Desire to excrete when prevented by any cause;

In the act of satisfying desires pleasure is apparent;

Hunger is appeased;

Movement is unconstrained;

Emunctories are unobstructed and excretion is active.

---

* "Origin of the Will." *Penn Monthly* for June, 1877, p. 446.

All pains and pleasures are relative and intense in proportion to the precedence of one or the other extreme.

The pangs of parturition are obstructive excretory, and what obstetrician has not noticed the happiness of accomplishment by the mother?

A pleasure is often due to the pre-existence of pain, and bearing upon the evolution of the reproductive excretory cellular into a desire, which, in its influence upon animal life, is second only to that of hunger; this relativity must be borne in mind. The pleasurable anticipation of eating is a memory, the physical basis of which in the amœba is a motion of the molecules involved in assimilation; their activity, their tension (the hungry amœba is always more active than when fed). The reproductive excretory is in the amœba scarcely to be called a desire, so dependent is it upon the performance of the assimilative act. The desire is invoked in exact proportion to growth from assimilation, provided other means of consumption of this growth are not operative.

This is obvious throughout all animal life. When hunger is extreme the sexual desire is absent.

Full meals sometimes excite voluptuous feeling. The repression of this excretory desire for a time becomes painful, until readjustment enables vicariation.

The desire to excrete the sperm cell is the male peculiarity, the desire, when present, of the female being, as shown in the *Science* article, identical with hunger. It is the hunger of the ova, which are part of the female, and which by differentiation have come to be capable of satisfaction in the manners to which they have grown.

From the Synamœba stage in its denser envelope, preventing the escape of the cells for a longer period, this sexual excretory desire would increase. Differentiation of the sexual

hunger from the general hunger is shown in the Drysdale and
Dallinger Monad.   C. M. Hollingworth,* on the "Theory of
Sex and Sexual Genesis," assigns causes determining sex.
"Since germ cells are large and sperm cells are small, it may be
at once inferred that where they are found in different parts of
the organism the parts in which germ cells or their producing
organs are formed must be parts in which the conditions are
especially favorable to nutrition; and that the parts in which
sperm cells or their producing organs are found, must be relatively unfavorable to nutrition and favorable to cell division."

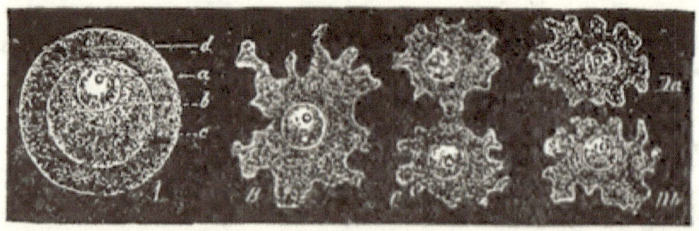

Fig. 2.   *Amœba sphærococus*, greatly magnified. A fresh water amœba without a contractile vacuole.   *A.* The enclosed amœba in the state of a globular
lump of plasma (*c*) enclosing a kernel and a kernel speck, (*a*) nucleus and
nucleolus.   The simple cell is surrounded by a cyst or cell membrane (*d*).
B. The free amœba, which has burst and left the cyst or cell membrane.   *C.* It
begins to divide by its kernel parting into two kernels, and the cell substance
between the two contracting.   *D.* The division is completed, and the cell substance has entirely separated into two bodies (*D a* and *D b*.)— *From* HAECKEL.

"The hypothesis is that a relative preponderance of the conditions on which cell division depends, causes the formation of
the female or male generative organs or determines the sex of
the individual."   Extending this to the amœba, the pure relativity of sex is seen.   If the amœba had undergone differentiation above some form by which it was engulfed, it could be
regarded as the male.   If it swallowed a synamœba then it is
the female cell, and the product of this sexual eating would be
either male or female, synamœba or amœba, according to the

* American Naturalist, July and August, 1884.

preponderance of differentiating influence or the disposition to increase by fission in the resulting fused mass.

Desires consisting of atomic tensions or affinities, the conditions of continuance or satisfaction of desire, involves feeling or sensation, a low form of consciousness. This is justified in considering our developed similar states during the same process of hungering, eating, etc., and, as in us, repletion discontinues desire; so does it in the lowest form of life we are discussing.

The sensations involved in assimilation would be difficult to separate from those concerned in pseudopodia protrusion or general locomotion, as they are identical in effect in the amœba. Admitting this identity, it is easy to see how by invagination of the ectoderm the later differentiation could occur by an enteric tactile developing in one direction, while the ectodermal would change with direct reference to locomotion or prehension. But, as even when the enteron is formed, a prehensile tactile sense is retained and developed, analogies between the nerve distributions to external and internal parts remain, though the sensations in many respects differ. The passage of materials in the intestines awaken few feelings so long as the adjustment is not disturbed, on the same principle that we do not feel external ordinary stimuli perpetually recurring.

Pressure is the feeling of constraint, interference with molecular and mass movement; it is a painful state of consciousness arising from the inhibited movement, the desire to move being consequent upon proper assimilation, and is referred to an interference with that function.

The hunger pain and appeasing hunger pleasure are due to and consist in chemical tensions and release from tension, the absence and presence of certain molecules.

This carried up the scale of metazoa convinces us that

desire, feeling, sensation, reside in every living cell in the body, and are not seated exclusively in nerve tissue. With the differentiation of function there will proceed changes of degrees of intensity of certain feelings in those living cells, but the fundamental hunger pain, and pleasure of its gratification, are never differentiated out of existence in any cell.

Desires, feelings, sensations, consciousness, cognitions, ideas, memories, emotions, etc., are, one and all, *conditions* of the molecules of the cells; and in the ravenous though unavailing appetite of some diseases wherein nutrition is at fault, the feeling is shown not to be solely located in the intestines, but all over the body, and the inability of physiologists to locate centers for desires in the brain is explained.

Whenever the exhibition of a feeling, or a feeling itself has been destroyed through injury, it has been through failure of the *tracts* which convey molecular movements generated by such feelings from the non-nervous bodily cells wherein these feelings are highly developed. The nerves are pure association systems, and where the feeling aroused in an organ has become, through constant repetition, associated with certain other feelings or with a motor expression, then nerves of association would be built up through least resistant lines. The organism consisting in the sum total of the life activities of its cells, the dissociation of the organism from its locomotory organs, the legs, cuts off the ability to walk, and paralysis of strands leading to the legs dissociates similarly.

Cutting off the organ of special sense or destroying its tracts similarly dissociate. There is a difference between cells acting for themselves or acting unitedly with others.

Returning to our amœba, the mobile granules and molecules moved with every impulse. Its sensations were motions and its motions sensations, the two were inseparable. With a change

in the density of its ectosarc retarding fission the morula form
arose, and in the break of the envelope amœbæ are, as might
be expected, liberated, but
they have inherited this
molecular development of
ectosarc induration and develop into synamœbæ as
did its parent.

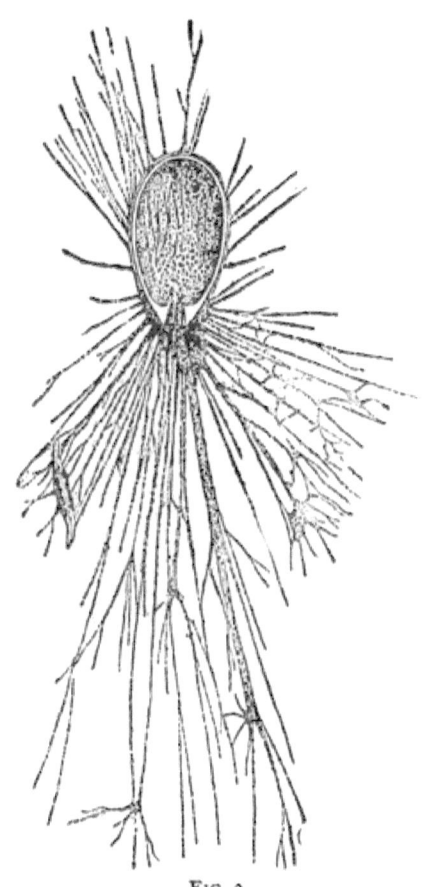

The planeæda developed
cilia through a similar law.
Owing to the difficulty of
withdrawing pseudopodia
once protruded, these atrophied into vibratile organs
through starvation, and the
hunger motions of the cells
set up oscillations of the
cilia, which, subserving the
life purposes better, were
perpetuated and the mo-
tions of the cells adjusted
themselves to the neces-
sity and brought the environment food to itself by
causing eddies, and a new
means of locomotion arose.

FIG. 3.

Gromia oviformis, with its pseudopodia
extended. From Carpenter's Microscopy.

Still, sensation and motion were identical, for the molecular
movements constituting sensation ended in their development of motions into gross locomotory motion, except
that the cilia were organs of locomotion while the body
remained sensitive. We may regard the cilium as formed
dead material in the main. Dallinger and Drysdales'

monad rejected it in eating its companion. This would
shadow forth the possibilities of a set of ciliary vibrations becoming known to the animal as tactile locomotory,
differing from, though ministering to, hunger sense, and a
change in the aggregation of the cell granules would follow.
When the cell granules moved in keeping with the ciliary motions either the locomotory memory or act was aroused. If in
accordance with hunger movements, then the memory of
hunger was aroused and this could react upon the cilia to move
them. When the ectoderm reached a stage of hardening admitting of no more strain upon it the central contents transuded
by some means, probably temporary rupture. The gaseous and
watery contents escaped and the animal collapsed into the
gastræa stage. The primitive enteric cilia still remain in the

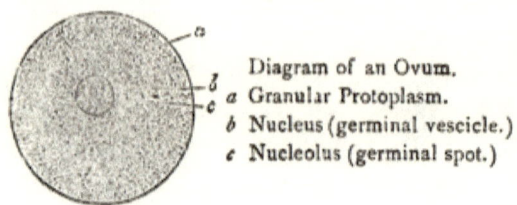

Diagram of an Ovum.
a Granular Protoplasm.
b Nucleus (germinal vescicle.)
c Nucleolus (germinal spot.)

Fig. 4.

respiratory tracts
as originally devel-
oped, but of course
changes between
ectodermal and en-
dodermal experi-
ences would differentiate the two areas, enteric cilia would
be useless and columnar epithelium would appear. The
inner remained subject to constant encounters, or nearly
so, and the outer had the brunt of every change. The
acœlomatous turbellaria appear to me to be more of
an aberrent type not in our phyllum; many of the forms
have undergone much development. The delamination origin of the gastrula stage could be one of those ontogenetic short-cuts often made in copying phylogenesis, the
ends attained being the same.

The scolecida or soft worm presents the most evident progress toward development in the vertebrate direction. Its

cœlom contains the first nutrient fluid allied to blood, but its circulation is not established. The denudation of the useless external cilia, though occasionally developed into stiff setæ in a turbelarian, follow the locomotory process changing to that characteristic of worms: elongation and contraction of the body length. Hubrecht's newly discovered worm, the *Pseudonematon,* illustrates this movement through the alternate contraction of longitudinal and circular muscles with a plexiform nervous system between. The motion is in some respects similar to the flow of amœba granules forward into an arm, but organization has restricted this to a to and fro motion. The shape of the body rendering this the easiest mode of progress, the sum of the life activities act in least resistant lines to enlongate and then contract the worm. Cause and effect exchange places in the circular fibers being placed through the elongation of motion in a state of inertia, enabling the contraction of the longitudinal with consequent adjustment of the circular fiber molecules so they contract to advantage. The repetition of these two opposed motions through the confinement of the skin rendering them about the only ones that could be made, initiated by the attractive affinities of the protoplasm, finally developed the contractile tissue.

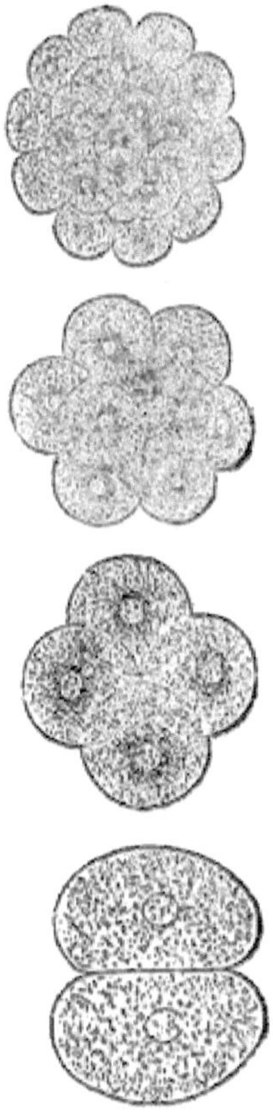

Fig. 5.—Various stages of the so-called cleavage process (Division of the Egg). From Gegenbaur. The first resembling the splitting of an amœba, and the last the morula or mulberry stage, corresponding to the synamœba.

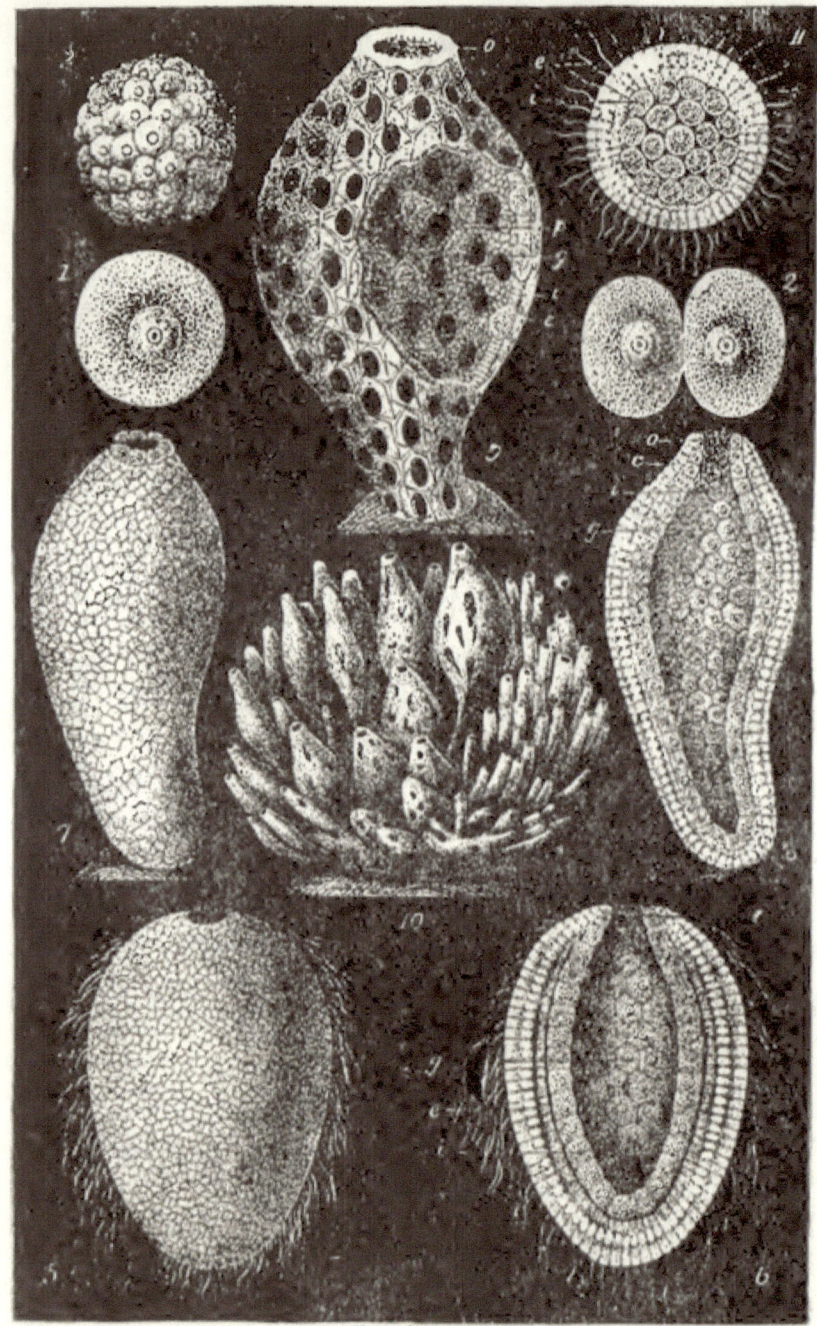

FIG. 6

Development of a calcareous sponge. *Olynthus.* From Haeckel. Showing the various stages passed through. Ovum (1), Morula (3), Planula (11), Gastrula (6, 7, 8, 9, 10).

The ventral location of the nervous system in the errantia or wandering worms, and others, is due to development through use of that region differentiating locomotor areas from the epidermis, and in insecta it is the persistence of the phylogenetic origin.

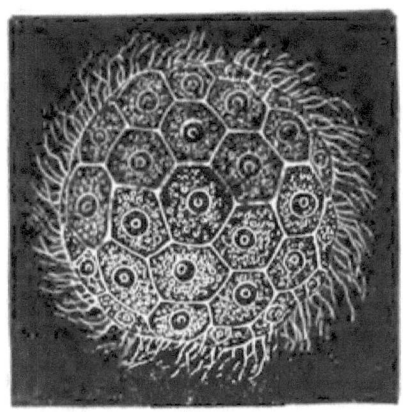

FIG. 7.

The Norwegian Flimmerball, (Magospaæra Planula) swimming by means of its vibratile fringes, as seen from the surface. From Haeckel.

In the worm stage the external tactile becomes fully developed through the heterogeneity of molar vibrations to which it is subjected.

One method of locomotion being possible another is also possible; differences or variations in one or the other, such as could be caused by mechanical means, could result in a sinuous movement arising from want of rigidity in the worm length. A complex of causes simultaneous and successive operate to change the usual mode of locomotion and introduce so-called compound reflexes.

The formed tissue of which Beale speaks is often excrementitious, and has through being useful been retained by the cells. The sandy covering of the rhizopod *Astrodiscus aræn-aceus* may have been "selected" by agglutination of the envelope with the particles, or the shell of a mollusc may form through excretory processes, or a covering may be acquired by squatter right, as with the hermit crab. It matters little to the animal. The fighting cock will use the steel gaffs with as much gusto as though they had grown from his legs, nor is the cell a particle more particular. If it find in its environment matter with peculiar properties it will through " selection " eat what it can and excrete the rest. If the excreted material have enough affinity for the cell to remain in its vicinity, and a life process is subserved by that fact, things chemical and mechanical in nature will conspire to associate the

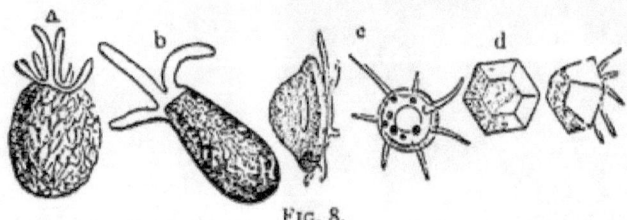

FIG. 8.

Testaceous forms of amœba Rhizopods :   A, *Difflugia proteiformis*; B, *Difflugia oblonga* ; C, *Arcella accuminata* ; D, *Arcella dentata*.   From Carpenter's Microscopy.   To show fortuitous coverings of rhizopods.

material with the cell. I regard nerve granules, such as are found arranging themselves, or being arranged, into, first, plexuses of fibers, and then definite tracts, as having arisen accidentally. As the rhizopod could not have acquired his overcoat where there was no sand, the ancestral worm which picked up a nervous system could not have done so in the absence of assimilable phosphates. The resulting nervous system became more and more definite in tract formation as motions became more definite between parts. These nerve granules had a molecular mode of action altogether different from anything experienced before by the animals. In higher forms the cell substance, which had the particular ability to excrete or secrete it (relative terms), formed along the area of the plexus and tracts; next an encapsulating membrane formed about it, in obedience to ordinary physiological and pathological processes, that an intermediary substance will be attracted and form around tissues, or even foreign substances, as a resultant of the mode of operation of the two tissues. In due time an area of nerve granule generation finds in itself small plexiform areas, which are turning points of direction for the molecular nerve discharge; and by encapsulating these the nerve cell is formed, which I regard as having no other function than a histogenetic one, aside from the molecular impacts passing through it. From protoplasm exuding the nerve granules, the nerve cells develop to that office. From indifferent tissue forming cartilage some of the latter form osteal cells.

A plexiform rudimentary nervous system conveys irritations over the body. When the discharges become definite the linear arrangement appears, as in ascidian embryo; the head end developing through tactile and rudimentary sense organs determining there with frontal impact of environment. Most influences acting to excite the squirt or vermicular motions from mouth to anus, a method of locomotion and ingestion at the same time.

CHAPTER III.

ORGANOGENY.

It is easy to see how the mechanical perforation of a cœlenterate sac caused the enteron to be completed, but the circulatory system origin is not so evident. The enterocœle is in direct communication with the enteron in cœlenterates, and the fluid it contains, as Huxley says, " represents blood." It is nutrient. The lacunæ of some worms are the next step toward a blood vascular system. The pseud-hæmal system of the annelida contains a substance resembling hæmoglobin; and with these facts before us we may construct the vascular system and its workings in some such way as this: The cœlom is a receptacle for a fluid containing nutrient matter which had strained through the endodermal cells and through interstices between them. Next the appearance of hæmoglobin or its equivalent in the cœlom; the peculiar properties of this substance consisted mainly in its solubility by alkaline fluids and its affinity for oxygen, "which is linked to it by ties so easily broken that it can be transferred to other easily oxidizable bodies existing by its side. It can be given up when its solutions are gently heated *in vacuo* or agitated at moderate temperatures with large quantities of inactive gases, such as nitrogen or hydrogen."* This oxygen carrier next found a cell especially adapted to its transportation.

The red globule being attracted from tissues where it received oxygen to those needing it, built up (by the identical processes which formed the nervous system) definite passages

* Gamgee's Phys. Chem. of the Animal Body, p. 91.

or canals from indefinite lacunæ as soon as definite attractions and oxygen furnishing areas arose through natural selection; for the same laws which govern the complete organism, save, develop, and kill off cells.

In discussing the cause of the rythmic contraction of the amœba we surmised for good reasons it was the discharge of CO_2 and water, owing to the ab.orption of oxygen. The direct union of the oxygen causing molecular increase of bulk in the sarcode. As the oxygen is carried along the blood vessels similar rythmic pulsations are induced by the cell movements in the abstraction of oxygen. Muscular cells form around these vessels as a consequence of the definite contractions thus induced. In veins through which the oxygen does not flow, this would not occur. Neither contractions nor development of muscle fibers. At first these motions would be feeble, but areas such as those in the Amphioxine vessels, where more oxygen had been absorbed, either through the crowding together of vessels as the animal shortened its length (this also causes the folding of the intestines) or through a physiological aneurism forming there. It is well known how the heart forms by the twisting upon itself of the vessel which forms the aorta. Haeckel [*] thinks that because ascidia have hearts and amphioxus none, that the latter lost it by reversion and we inherited it through the original pharyngobranchii. It is easier to suppose that the differentiation occurred in a later stage than amphioxus and that the primitive ascidian had no heart.

As to the respiratory development, this was originally through the whole skin then through the intestinal tract as in all cœlenterates and some fishes; next the internal and external gills develop more toward the head and with the adaptation of the perennibranchiate to land life, and in some forms

---

[*] Schöpfungsgeschicte, Vol. II.

before, the swimming bladder develops the air cells and becomes a lung.

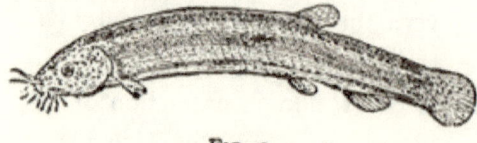

FIG. 9.

Cobitus fossilis. It swallows air bubbles, which pass through the intestines, where the mucous membrane takes up the oxygen for respiration.

In the June 25, 1881, Science, (N. Y.), I published the following:

"There are many reasons for believing that the thyroid and thymus are rudimentary gills, one of the main objections to the view being the structure of these bodies; but in the light of modern biology structure is almost meaningless in homologizing. Besides the tissues of these parts are not the same in all animals. Owen, Vol. I, p. 565, says the thymus appears in vertebrates with the establishment of lungs as the main or exclusive respiratory organ. In Siren and Proteus the thymus is wanting, as in all fishes. Gegenbaur, Comp. Anat. p. 554, speaks of the thyroid as an organ with unknown physiological relations, and that ' in fishes it is placed not far from the point at which it was formed, that is, at the anterior end of the trunk of the branchial artery and between it and the copula of the hyoid arch ; in amphibia near the larynx, and is set on the inner surface of the posterior cornua of the hyoid.' Gegenbaur considers it an organ of use among Tunicata. This latter idea, as well as the one I have advanced, needs verification. I am unwilling to devote more time to the subject until I can ascertain whether some one has not preceded me in announcing the homology, if it be one. Much light can be thrown upon the disease known as goitre by clearing up this point."

Fatty degeneration or conversion into another tissue is the general fate of the thyroid and thymus. Since the above was published I am led to include the tonsils in this derivation. "The fœtal lung structure resembles the thymus, and the pleura covers it."*

* Klein's Atlas of Histology, p. 244.

The osmotic influence in respiration I regard as osmosis generally should be, depending mainly upon the conditions existing on both sides of the membrane, rather than as having reference to any mysterious property of the membrane itself aside from a peculiar adjusted permeability.

This is the developed trait and structure of the air cell. The hæmoglobin is the active agent in effecting respiration.

The lymphatic system, Rindfleisch considers as a receiver of materials in excess of nutrition; hence when clogged as by mercury, etc., there will be luxurious growths, neoplasms and catarrhs. It may be likened to the original cœlomatous cavity, as the receptacle of extra pabulum, and whenever lymphatic hearts exist in an animal, oxygen must have been conveyed through them, if my theory be true, and the defective oxygenation of the lymph in man and such other animals as have not these hearts point to the reason for the arterial pulsations existing without similar movements in other channels. The intestinal rythm partakes of this nature. Much oxygen is swallowed with food and abstracted directly by the intestines from it, hence the vermicular motions; but there is a mechanical tactile influence also operative there. This is the intestinal locomotory tactile activity perpetrated from the ectodermal origin.

A lack of oxygen excites vermicular motion, just as the activity of the amœba is excited when it is hungry. The complete absence of oxygen from both amœba and intestine would paralyze the motions. Excess of CO_2 excites expulsory motions in both, the excretory presence having developed to expel this gas, even with the absence of much oxygen. When the blood is saturated with oxygen apnœa follows. The plethora of the molecules stay all motions in both cell and higher organism or organ.

3

"The size of the red blood corpuscle is greater in the ratio of the persistence of the branchial apparatus, and the perenni-branchiates or deciduous gilled animals present the biggest blood discs absolutely, as well as in proportion to the size of the body, of all vertebrate animals." *   The blood corpuscles are relatively smaller in both sides of the vertebrata, showing that as the gills become less effective and before the lungs were fully developed the increase in the blood disc was of a vicariating nature.   Prof. Harrison Allen suggested to me that this would admit of more oxygen to the disc and probably enable it to be held longer.   The idea is a good one.   After the lungs are established the corpuscle grows small again, and this fact, with due consideration of third causes, which *always* operate, will work out the problem of red blood corpuscle sizes throughout animal life.

Oxygen affinity, from amœba up, seems to be the prime direct agent in movement.   Of course other nutrient processes must be considered, but they operate less vehemently and tend to facilitate the establishment of oxygen consumption.   Given an environment in which the organic matter finds easy entrance to protoplasm, the activity of the cell or cells increases with the added facilities for its assimilation of pabulum and at the same time the oxygen consumption increases with attendant activity.   The CO_2 evolution is in direct proportion and dependent upon this activity.   Where, from an irregular means of obtaining oxygen, only erratic, here and there, cellular action could result, with the passage to and fro of oxygen carriers, rythmic to and fro motions of rudimentary muscle cells would develop, and where, as in the ascidian, this rythm became fixed in a mouth to anus direction, the tendency toward fixation of a method of supply of oxygen would be a consequence. This appears to involve the constant definite supply of oxygen to sphincter muscles while in tonic contraction, and the establishment of subsidary arrangements in the way of blood supply to produce it. Muscle cells would differentiate with abilities as widely unlike as those of unstriped and striped aggregates. The striped have been evolved from the unstriped. The former being under the control of the will and acting more promptly over large extents than the unstriped. Abilities which may be due to arrangements of the muscle cells themselves, to the development of these cells, and to associated apparatus. The stimuli rates necessary to contract red and pale striated muscles are ten in the former to twenty or thirty in the latter per second, showing that with blood increase there is more prolonged action for the same stimulus. The contraction is quicker in insects than frogs; in the heart muscle than in the intestinal smooth muscle; the tendency of differentiation being to prolong the contraction for a similar stimulation. The translocation of molecules are in rectangular directions. In the extension of an amœbic part by attraction (of oxygen), there is consequent contraction of the mass across the line of extension; so that contraction is secondary and a result of the propulsion of particles apart. In the expulsion of the vacuole contents there is an aggregation in general of the mass. From this and other phenomena, I judge that the undifferentiated cell which would expand with the introduction of oxygen, at least until the atomic interchange had been effected, might, as CO_2 was evolved as a waste product, then slightly contract. But with the differentiation of muscle cells the aggregative tendency of the cell contents implied the presence of some probably crystallizable substance, *particles of which when oxidized* would be drawn together as long as the temporary oxidation lasted. Such a property might be developed from hæmoglobin acquiring attractive qualities when situated in muscle precincts. In fact, hæmoglobin, with its peculiar faculty of loosely holding oxygen, could be thus developed when situated axially in a cell which constantly tended to act in a certain way, as the muscle cell does. A bar of iron, if hung in the line of the magnetic meridian, or subjected to frequent concussions, will acquire magnetic properties. It is the constant operation of the law of adjustment to circumstances. A cell, a tissue, or an animal which acts in a certain way finds that chance throws in its path materials of which it can make use; and even these undergo adaptation to the constant influence. This action and reaction has built up the muscle complex. That oxygen is a factor in the contraction is not disproven under circumstances which apparently, but do not really, cut off that gas. Englemann* has shown that muscle in atmospheres of carbonic acid, carbonic oxide, hydrogen, etc., soon loses its irritability, but in oxygen preserves it for a long time. *Apropos* of the tonic contraction implying a modified vascular distribution in muscles subject to it, " Ranvier demonstrated a peculiar condition of the minute veins and capillaries in the red muscles of the rabbit, these vessels being possessed of sinuous and spindle-shaped dilatations, owing probably to the almost permanent contraction of these muscles." †

The parallel direction of the capillaries with the muscle fibers is also significant.

Certain medicines, like ergot, may influence this oxygenating function and thus exert their characteristic effects.

The evolution of the liver is through certain scattered enteric cells developing an ability to elaborate certain substances.

---

* Pflüger's Archiv. II, 1869, 243.

† E. Meyer, quoted by Klein, Atlas of Histology, p. 80.

Natural selection grouped those cells in areas where this substance was most likely to be encountered, and hence the organogeny can be traced. That bile facilitates the intestinal operations in my opinion points to an adjustment of the enteron *for* that excrementitious substance rather than that the liver was made *to* secrete bile to act on the canal. The liver cells have an elaborating function of a peculiar kind allied to the general assimilative, and doubtless other functions than the original have been added to the liver in its development.

The relativity of the terms excretion and secretion is apparent in the complementary functions developed by cells. In the greedy absorption altruism or generosity is absolutely absent, and what one cell cannot take up goes to another. In cell excretion substances are changed more or less, and a contiguous cell *must*, if situated so it cannot reach the food direct, put up with such excreted matter. It adjusts itself to that sort of food and in time cannot survive on any other. Due consideration of this will account for the multitude of chemical combinations in the fluids and tissues of the body.

An ethological deduction is not out of place here. Each ciliated cell whips the environment to bring itself food ; it cares nothing for its neighbor, nor is it grateful to the preceding cells which aided it in starting the circulation. This egoism or selfishness is that of the chemical elements which grab where they can, but in high development the organism sees that others are benefited by its egoistic acts and instantly deliberates how to turn this to account. It suggests that as its neighbor has derived an advantage from this act some sort of recompense is due, and where nothing more substantial is obtainable *gratitude* is demanded, which is merely a generalization for the disposition to requite favors in the future ; to pay, when the opportunity arises.   Our reprehension of ingratitude is in the abhorrence of not granting the *quid pro quo.*

The adjustment of cells to certain media may entail the necessity for the continuance of the presence of substances even after having left the medium in which the adjustment was acquired.   For instance, the desire for salt *with* food in excess of that furnished by the food, indicates the piscine adjustment ; and the presence of salines in the fluids evidence the omniprevalence of this cell adjustment.   Some savages and animals have acquired a disrelish for salt through circumstances developing that disrelish, such as absence of the article from the country lived in, or ignorance of the mineral as found on land partaking of the nature of that to which the marine ancestor was accustomed.   Nevertheless a modicum is always taken in with the vegetable or animal food.

Some of the changes wrought by the erect posture following the previous quadrupedal, were noted by me in the subjoined which was published in the *American Naturalist* January, 1884. The substance of the article entitled " Disadvantages of the Upright Position," I read before the University Club of Chicago, April 18, 1882:

The immediate and remote causes of things have been and will be sought by thinkers who are not afraid to follow wherever facts lead them.   The doctrine that there is no effect without an antecedent cause, has met with fierce opposition from those who saw that the logical conclusions of correlated facts, such as are presented by Darwin, tended to the overthrow of puerile legends they believed in, and who were content to imagine that everything was causeless, or, at best, originated in some inscrutable way.   The Arab, upon having the sidereal motions explained to him, said, " You trouble yourself greatly about things not intended for you to know.   Even though what you tell me is true, the Koran leads us to believe otherwise. Mohammed taught us sufficient, and his followers can torture you out of your rationalism. Forbear your heretical facts!"

The mechanical nature of things animate is as old in theory as Democritus, 500 B. C.; and Giordano Bruno, in A. D. 1600, for having amplified the Democritic idea, was burned at the stake. Kant granted a mechanical cosmogony, but in organic nature claimed *causæ finales.* The battle of *causæ efficientes* was fully won by Galileo, Copernicus, Kepler, Newton, Herschel, Laplace, etc., so far as the inanimate universe was concerned, but the mechanical conception of that which pertains to living things was hinted at by Aristotle. Geoffrey de St. Hilaire contended against Cuvier for the mutability of species and the monistic theory. Treviranus, Oken, Goethe, Lamarck, and, in our day, Darwin, Haeckel, Huxley, have carried on the warfare. Herbert Spencer advanced a mechanical physiology and morphology. His has carried the conception into histology, and Cope into palæontology. The unity of the laws which control organic and inorganic nature are to-day fully recognized by those who stand in the front rank of investigators and thinkers, but not until completer textbooks from the new standpoint shall have found their way into the hands of medical students and naturalists generally, will common recognition of the success of the mechanical idea be obtained.

Assuredly, the teleological is a very lazy way of thinking. It amounts to taking things for granted as so, because they are so. It bars all inquiry, stops all investigation, and hands us, bound hand and foot, to ignorance and superstition.

Mechanical influences, such as impacts and strains, permanently altering animal organs have been discussed by Professor E. D. Cope in the *American Naturalist,* in articles entitled, Origin of the Foot Structures of Ungulates, April, 1881; Effects of Impacts and Strains on the Feet of Mammalia, July, 1881; by Alpheus Hyatt, Transformations of Planorbis at Steinheim, with Remarks on the Effects of Gravity upon the forms of Shells and Animals, June, 1882. In articles published in the January and February, 1881, numbers, I attempted a disquisition upon physical influences in their relations to comparative neurology, and in the July, 1881, number of the *American Naturalist*, On the Origin and Descent of the Human Brain, pointed out some hitherto neglected mechanical factors in the development of the organ of the mind and its osseous envelope.

While engaged in anatomical studies, the idea that there was a definite reason for everything, and that we might some day discover the reasons for many things not now known, was ever present to my mind. I could get half lights and glimpses of causes from hints in Henle, Holden, or Sharpey and Quain, and fancied I saw matters clearly enough in some particulars, only to be confused by contradictory experiences subsequently.

There seemed to be a definite enough law in the formation of valves in the veins, for instance, but every student was compelled to learn the location of these valves by arbitrary exercise of the memory. I think that every student will conclude at the end of this paper that it is easy enough *now* to remember which veins are valved and which are not. Let me present the subject just as it perplexed me at first Nothing could be simpler from the teleological standpoint, than that we should have valves in the veins of the arms and legs to assist the return of the blood to the heart against gravitation, but what earthly use has a man for valves in the intercostal veins which carry blood almost horizontally backward to the azygos veins? When recumbent, these valves are an actual detriment to the free flow of blood. The inferior thyroid veins which drop their blood into the innominate are obstructed by valves at their junction. Two pairs of valves are situated in the external jugular and another pair in the internal jugular, but in recognition of their uselessness they do not prevent regurgitation of blood nor liquids from passing upwards.

An apparent anomoly exists in the absence of valves from parts where they are most needed, such as in the venæ cava, spinal, iliac, hæmorrhoidal and portal. The azygos veins have imperfect valves.

Place man upon "all fours" and the law governing the presence and absence of valves is at once apparent, applicable, so far as I have been able to ascertain, to all quadrupedal and quadrumanous animals : *Dorsad veins are valved; cephalad, ventrad and caudad veins have no valves.* The apparent exceptions to this rule, I think, can be disposed of by considering the jugular valves as obsolescing, rendered rudimentary in man by the erect head, which in the lemur stage depended. The rudimentary azygos valves may be a recent creation, and an explanation of their presence may be found in the mutability of the cardinal system. The azygos veins are derived from the intercostal, and the rudimentary valves may be a remnant of the original condition. The single Eustachian valve, being large in the fœtus, has a phylogenetic value.

The only reason I can assign for the absence of cephalic and cervical valves generally, while the jugulars possess them, is, that the jugular system was the most important to our quadrupedal ancestors with dependent heads, hence valves developed in them, and owing to the cranial blood vessels developing, *pari passu*, with the cranium and its contents generally, largely after man had assumed the erect position, the valvular formation elsewhere in the head would not occur while the jugular valves became rudimentary.

Certainly valves in the hæmorrhoidal veins would be out of place in quadrupeds, but to their absence in man many a life has been and will be sacrificed, to say nothing of the discomfort and distress occasioned by the engorgement known as piles, which

a, refers to the spinal system ; *b*, jugular and caval to femoral ; *c*, brachial ; *d*, intercostal.

the presence of valves in these veins would obviate. The spermatic valves are as useful in man as in other animals.

A glance at the accompanying diagram will afford an idea of the confusing distribution of valved and unvalved veins in the human being.

The position assumed by these valved veins when man is placed on all fours, corresponds with those to be found in quadrupeds, thus:

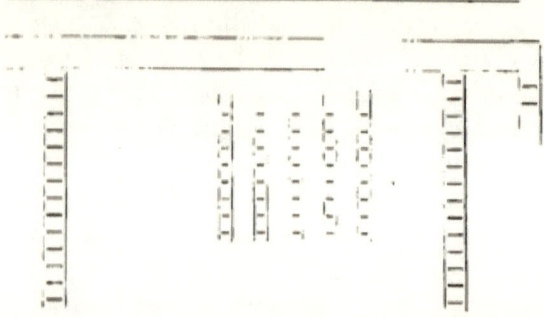

A noticeable departure from the rule obtaining in the vascular system of Mammalia also occurs in the exposed situation of the femoral artery in man. The arteries lie deeper than the veins, or are otherwise protected for the purpose, the teleologist would say, of preventing hæmorrhage by superficial cuts. From the evolutionary standpoint it would appear that only animals with deeply-placed arteries would survive and transmit their peculiarities to their offspring, as the ordinary abrasions to which all animals are subject, not to mention their fierce onslaughts upon one another, would quickly kill off animals with superficially located arteries. But when man assumed the upright posture, the femoral artery, which was placed out of reach on the inner part of the thigh, became exposed, and were it not that this defect is nearly fully atoned for by his ability to protect the exposed artery in ways the brute could not, he too would have become extinct. Even as it is, this aberration is a fruitful cause of trouble and death.

Another disadvantage which occurs in the upright position

of man, is his greater liability to inguinal hernia. Quadrupeds have the main weight of abdominal viscera supported by ribs and strong pectoral and abdominal muscles. The weakest part of the latter group of muscles is in the region of Poupart's ligament, above the groin. Inguinal hernia is rare in other vertebrates because this weak part is relieved of the visceral stress, but as the pelvis receives the intestinal load in man, an immense number of trusses are manufactured to supplement this deficiency. It has been estimated that twenty per cent. of the human family suffer in this way, and strangulated hernia frequently occasions death.

If man has always been erect from creation, then we have nothing to hope from the future by way of an alteration of this defect. The same percentage of humanity will suffer to the end of time; but considered mechanically the so-called conservative influence of nature which will tend to pile up additional muscular tissue in this region by reason of the increased blood supply to that part, aided by natural and sexual selection, will eventually reduce the percentage of ruptures greatly, if it does not finally correct the trouble altogether. The liability to femoral hernia is similarly increased by the upright position.

The peritoneal ligaments of the uterus subserve suspensorial functions in quadrupeds fully, which requires much ingenious speculation to be faintly seen in man. The anterior, posterior and lateral ligaments are mainly concerned in preventing the gravid uterus from pitching too far toward the diaphragm of four-footed animals. The round ligaments are absolutely meaningless in the human female, but in lower animals serve the same purpose as the other ligaments. Prolapses uteri, by the erect position and absence of support fitted to that attitude, are thus rendered frequent, to the destruction of health and happiness of multitudes.

As a deduction from mechanical laws, it could easily be imagined that an animal or race of men which had the longest maintained the erect position would have straighter abdomens, widely flared pelvic brims with contracted pelvic outlets, and that the weight of the spinal column would carry the sacrum lower down, and in general terms we find this to be the case. In quadrupeds, the box-shaped pelvis, which admits of easy parturition, prevails, but where the position of the animal is such as to throw the weight of the viscera into the pelvis, the brim necessarily widens, these weighty organs sink lower, and the heads of the femora, acting as fulcra, admit of the crest of the ilium being carried outward, while the lower part of the pelvis must be contracted. This box shape exists in the child's innominate bones, while its protruding abdomen resembles that of the gorilla. The gibbon exhibits this iliac expansion through the sitting posture, which developed his ischial callosities. Similarly iliac expansion occurs in the chimpanzee. The Megatherium had wide iliacal expansion, due to its semi-erect habits, but as its weight was mainly supported by the huge tail with femora resting in acetabula placed far forwards, the leverage necessary to contract the lower pelvis is absent. Professor Weber, of Bonn, noted by Carl Vogt, "*Vorlesungen über den Menschen,*" etc., distinguished four chief forms of the pelvis in man: the oval, round, square and cuneiform, owned in order by Europeans, native Americans, Mongols and black races. Resting upon its own merits as an osseous mechanical proposition, it would seem that the older the race the lower the sacrum and the greater the tendency to approximate the larger transverse diameter of the European female. The anteroposterior diameter of the simian pelvis is usually greater than the transverse; a similar condition affords the cuneiform, from which could be inferred that the erect position in the negro races had not been so long maintained as by the Mongols, whose pelvis assumed the quadrilateral shape owing to persistence of spinal axis weight through greater time; this pressure has finally culminated in pressing the sacrum of the European nearer the pubes, with consequent lateral expansions at the expense of the antero-posterior or conjugate. From Marsupialia to Lemuridæ the box shape pelvis persists, but with the wedge shape induced in man a remarkable phenomenon also occurs in the increased size of the fœtal head in disproportion to the contraction of the pelvic outlet. While the marsupial head is about one-sixth the size of the smallest part of the parturiant bony canal, the moment we pass to erect animals the greater relative increase is there in the cranial size with coexisting decrease in the area of the outlet. This altered condition of things has caused the death of millions of otherwise perfectly healthy and well-formed human mothers and children. The palæontologist might tell us if some such phenomenon of ischial approximation by natural mechanical causes has not caused the probable extinction of whole genera of vertebrates.

If we are to believe that for our original sin the pangs and labor at term were increased, and also believe in the disproportionate contraction of the pelvic space being an efficient cause of the same difficulties of parturition, the logical inference is inevitable that man's original sin consisted in his getting upon his hind legs.

Something of the changes noticed in the angle at which the head of the femur in set upon the shaft at different ages, is also noticeable phylogenetically. The neck of the femur in the child is obliquely placed, but in the adult is less so, and in advanced age tends to form a right angle with the socket. Both in the advance of age in the individual and the tendency of an animal to assume more and more the upright posture, this change of angle seems attributable to no other cause than bodily weight against the femoral heads.

This subject is not without direct application. Gynæcologists cause their patients to assume what is called the knee chest position, a prone one, for the purpose of restoring uteri to something near a natural position. Brown-Sequard recommends drawing away the blood from the spine in myelitis, or spinal congestion, by placing the patient on his abdomen or side with hands and feet somewhat dependent. The liability to spina-bifida is greatest in the human infant through the stress thrown upon the spine, and the absence of delivery troubles among lower races have reference to discrepancy between pelvic and cranial sizes not having been reached by those races. The Sandwich island mother has difficult delivery only when her progeny is half white, that breed being larger in the forehead than the native child.

Dr. Crichton Brown notes in West Riding Asylum Reports that a great number of insane cases originated through disasters directly due to cranial and pelvic discordances at child birth.

The mechanism of the body, when fully recognized as mechanism and nothing else, and as governed by mechanical laws, physical as well as chemical influences, will place forthcoming physiological studies upon a broader, safer foundation, and result in grand generalizations.   The hydro-dynamics of animal life would alone furnish a theme for thousands of investigators.   At present the world goes on in its blindness, apparently satisfied that everything is all right because it exists at all, ignorant of the evil consequences of apparently beneficent peculiarities, vaunting man's erectness and its advantages, while ignoring the disadvantages.   The observation that the lower the animal the more prolific, would eventuate the belief that the higher the animal the more difficulties encompass his development and propagation, and the cranio-pelvic incompatibility alone may settle the Malthusian doctrine effectually for the higher races of men through their extinction.

CHAPTER IV.

GENESIS.

Herbert Spencer* thus tabulates the relations existing among the different modes of genesis :

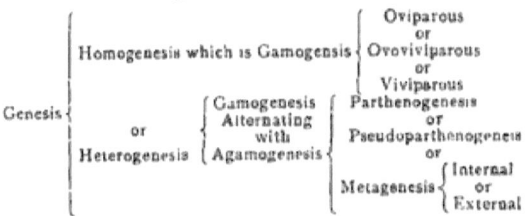

Spencer quotes (ibid) Professor Huxley's classification which makes every kind of genesis a mode of development.

```
                ┌ Continuous              ┌ Growth
                │                         ┤ Metamorphosis.
Development ┤                             └
                │ Discontinuous  ┌ Agamogenesis  ┌ Metagenesis
                └                ┤ Gamogenesis.  ┤ Parthenogenesis.
                                 └
```

The Spencerian and Huxleyian schemes may be put into simple language thus :

```
                      ┌ Similar to Parent form which is from  ┌ From the Egg
                      │           Sexual Union                │   (oviparous)
                      │ (Homogenesis)       (Gamogenesis)     │       or
                      │                                       │   From the Egg and
(Genesis)             │                                       │ bornalive
Breeding is through ┤            or         ┌ From sexual union │ (Ovoviviparous)
                      │ Different from       │  (Gamogenesis)    │       or
                      │  the Parent          ┤ Alternating with  │ Born alive
                      │ (Heterogenesis)      │ Breeding without Pre- │  (Viviparous)
                      │                      │ vious Sexual Union │   Born from regular
                      └                      └  (Agamogenesis)    │ generative apparatus
                                                                  │ without previous sex-
                                                                  │ ual union of parents
                                                                  │  (Parthenogenesis)
                                                                  │       or
                                                                  │ Similarly from non-
                                                                  │ differentiated gener-
                                                                  │ ative apparatus
                                                                  │ (Pseudo-Parthenogen-
                                                                  │ esis)
                                                                  │          or
                                                                  │          ┌ From the
                                                                  │          │  outside
                                                                  │ Budding ┤    or
                                                                  │          │ From the
                                                                  └          └  inside.
```

---

* Principles of Biology, Vol. I, p. 215.

Development
- Continuous.
  [As in the formation of a colony such as the sponge, the coral aggregates in the individual which is made up of a lot of cells bound together (and in a certain sense Society, Nations, the Earth and the Universe,)]
- Discontinuous.
  (As when the Amœba throws off and excretes individuals or individuals give birth to individuals.)

Growth (By mere quantitative increase of cells.)
    or
Metamorphosis (By radical change in the characteristic of the cells.)

- Without sexual union.
  - Metagenesis
  - Parthenogenesis (as explained above.)
- Through sexual union.

We considered the pure relativity of the matter of sex in low forms. This relativity could be extended to atomic combinations. The germ molecules differing from the sperm molecules through differentiation of the latter, the formation of H_2O may without much stretch of the imagination be taken to represent the preponderance of quantity in the hydrogen to represent the germ molecule, and in the heightened activity of oxygen the single but *relatively* more chemically important atom will represent the sperm cell. In the amœba simple fission is the starting point for animal life, and this being a product or consequence of growth, it matters nothing whether that growth was obtained by a cannabalistic act or by ordinary assimilation of other organic matter. The synamœba being developed, as hinted, naturally, by accidental hardening of the ectosarc, the rupture of the envelope liberates amœbiform offspring (at this stage heterogenetic) which subsequently develop into synamœbæ (then homogenetic.)

The amœba would be agamogenetic if its fission followed from growth either through assimilation of food or of another amœba. But suppose the swallowed amœba to differ from the cannibal. If inferiorly, *the product could not be otherwise than amœbæ.* If superiorly, the product could not be otherwise than a higher type of amœba or resemble its parent, because the differentiated tissue of the higher type would reappear in the offspring or would not reappear in some of them. Here, then, we have indifferently agamogenesis resulting always in homogenesis, and the same individual capable of the gamogenetic act (not differentiated from the assimilative act) developing through definite laws and for evident reasons either through homogenesis or heterogenesis.

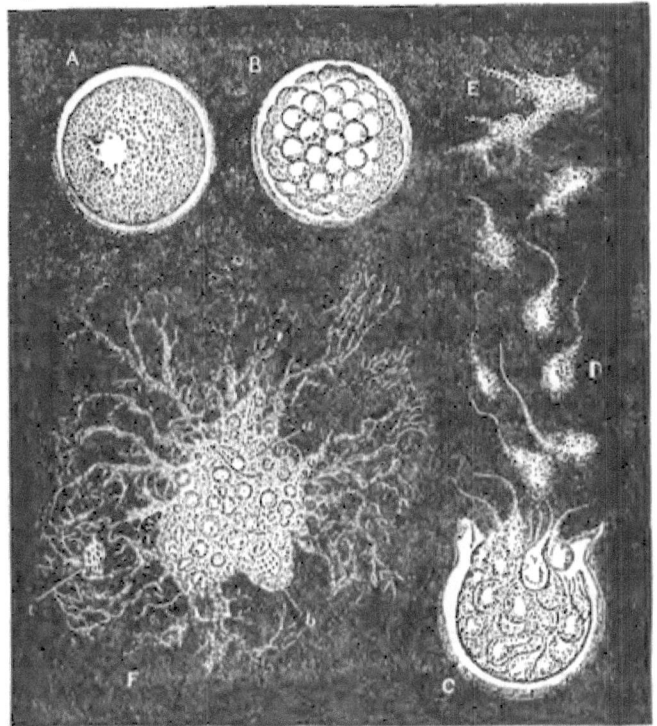

FIG. 10.

A, Encysted statospore; *B*, Incipient formation of the swarm spores, shown at *C* escaping from the cyst, at *D* swimming freely by their flagellate appendages, and at *E* creeping in the amœboid condition; *F*, Fully developed reticulate organism, showing numerous vacuoles, *a*, and captured prey, *b*, *c*. From Carpenter's Microscopy.

In the morula (synamœbic) form, which is at first heterogenetic and the offspring develop homogenetically, a teratological influence such as a reversion to the softer ectosarc of the amœbic ancestor would bring about "arrested development" and result in that instance in heterogenesis, which is the rule in all monstrosity production.

4

I incline to the belief that we are not to regard the planula and gastrula stages as otherwise than consequences of the synamœba stage, and hence the relationship of these stages are closer than might otherwise be supposed.

Metagenesis, which is usually agamogenesis, from no specialized part of the parent, is necessarily heterogenetic, for it involves a reversion toward the less perfect style of reproduction. Even though the gamogenetic act preceded the appearance of this method of heterogenesis, it had not influenced the product, and hence metagenesis may be said to be agamogenetic always. The external budding in most cases of metagenesis point to a faulty direction toward the paternal tissue and suggest a cause for the heterogenesis. Sexes being relative in very low stages, the ectoderm budding appears as a developed remnant of the stage when the organism was relatively a female and the endodermal less differentiated cell could not influence it in this higher stage. When metagenesis is internal it is a pure reversion, as might be expected in entozoa, such as the *Distoma*.

Parthenogenesis, which is agamogenesis carried on in a sexual organ, is the same thing, with the difference of the development of special cells differentiating from the general cell mass as best able to furnish both the ectodermal and endodermal peculiarities.   Von Siebold restricts this to the exclusion of a similar process, which Spencer designates as pseudo-parthenogenesis, and which Huxley claims is from pseud-ova and not true ova.   This is the stage intermediate between metagenesis and true parthenogenesis, and may be a reversion from the latter or a development from the former.   It is the transition period for the establishment of the higher.

With the advent of genesis by division of the amœba the product may be AAAA with a developmental reason for the occasional introduction of B.   Now this sort of reproduction may be brought about by a change in diet. A new form of amœba may originate from primitive though peculiar chemical conditions made possible by a change of environment, thus bringing about a temporary or permanent change of the tissues, and I doubt not many cases of species origination, such as the short legged sheep Darwin mentions and sub-varieties of animals originated embryologically. This phase of progress by natural selection acting upon the parent cell is afforded in the development of the queen bee through change of food. It shows that even in high forms the classification of Huxley is justified, and that means of development originally acting through the environment may still be brought to bear upon the ovum.

The step from AAA with fortuitous B. Homogenesis, with occasional heterogenesis in an evolutionary way, will be to BAAAA by reversion and with A developing subsequently like the parent. The product of relative male B by relative female A, would be either A or B, just as the cell elements of either parent predominated. The product of agamogenetic B would be either A or B, or A first developing into B. (The intermediate and lowest stages must be passed through.) The product of gamogenetic B with C would be either B or C, or, by reversion, A.

Metagenesis in A is impossible; in B would be identical with its usual mode of increase, *i. e.,* A, when agamogenetic, but when gamogenetic with C, the relative male. Then metagenesis by persistence of relative male attributes of ectoderm would produce B, the previous A stage being inevitable. The pathological dermoid cysts are explained by this.

Relative female C with relative male D could thus produce C or D, etc.

The occasional appearance of hermaphrodism can be accounted for by remembering that every product of A and B,

B and C, must have inherited some of the elements of both
The presence of rudimentary sexual organs in both male and
female attests this.   Where, by a fault, some of the organs
peculiar to the other sex have not atrophied but have maintained a certain stage of development, hemaphrodism is the result.   In speaking of the relativity of sex in organisms, as soon
as differentiation fixes the sexual organs so that intromission
is possible only *in one way*, this relativity gives way to *absolute*
sex, *though as regards other organs and even mental traits it
still holds good.*

Darwin's theory of pangenesis is worth repeating here as a
provisional hypothesis in explanation of heredity.   It assumes
that the various cells throughout the body throw off minute
granules called gemmules, which circulate freely through the
body and multiply by subdivision.   These gemmules collect
in the reproductive organs and products, or in buds, when that
mode of reproduction obtains; so that the egg or bud contains
gemmules from all parts of the parent or parents, which in development give rise to cells in the offspring similar to those
from which they were given off in the parent.   It is also assumed that these gemmules need not in all cases develop into
cells, but may be transmitted from generation to generation
without producing visible effect until a case of atavism appears.

The amœboid cell A may pass to a morula stage through ectosarcal change similar to that of the synamœba.   It will throw
off A and B cells, both containing attributes of the parents,
though in differing degrees; the essential change being peculiarities of the ectodermal forming cell acquired by an endodermal one.

Aa, Ab, Ba, Bb, may represent the preponderating influence
of the relative sexes.   Certainly the product of B is A, with a
tendency to form B, and in its embryo condition may be called

ovular, and as containing inherent molecular possibilities acquired from both parents, which enable it to take up from its environment the atoms for which it has affinities, and which develop it into the morula form B. If while in this form, quiescent through repletion, it be impregnated or it assimilate additionally C, the ciliated still higher differentiated cell, the planula stage of ontogenesis would result. A many celled mouthless larva would be the product through B dividing into A, which forms B cells, some of which still further develop into C. Whether this passes direct to the gastrula stage D, by invagination, or the gamogenetic union results from C and D in D, matters nothing. At this stage we can halt and review our ground, and consider Darwin's gemmule idea to have arisen from the necessity for compactness of formation, for no such stages as subsequently follow the gastrula, or for that matter precede it, have been discovered in higher metazoa.

Why does ontogenesis copy phylogenesis? is the question. We have seen that the amœba may develop directly agamogenetically, through acquiring molecular possibilities from its environment.

An amœbiform white blood corpuscle can also thus develop. Where its chemical affinities are adjusted to make the first step toward differentiation possible, the next step becomes also possible; and, as chemical union constitutes the main cause of development in the amœba, through selecting materials in its environment, the materials so selected will exist in the organism, and may be selected therefrom by the amœbiform cell as readily as from the ancestral environment—even better, for the bodily environment of the amœbiform corpuscle differs from that of its ancestor, and the former has undergone differentiation through this difference, the most noticeable being in its sluggish motions, probably due to its repletion, which in turn has developed a less easily pleased appetite (molecular affinities changed from that of the amœba, through molecular constitution differing slightly from it by development).

This difference is 'one of the causes of ontogenesis not copy-ing phylogenesis in every particular. In passing to the Ab stage (the potential morula) it is sufficient if the corpuscle ac-quire ectosarcal differentiation *tendency*, which under proper circumstances might develop dynamically. This, and the en-suing step, could be abbreviated by the ingestion of a ciliated or even an embryonal ciliated cell, or by its acquiring the po-tentials of what *would* develop a ciliated cell, were fission to take place. All this might occur and yet under the micro-scope the amœbiform cell would not apparently differ from its fellows. We know that these potencies exist in the bee ovum; the larva from which can be developed into a higher than the neuter form through change of food.

The 'potential gastrula stage is easily seen as a mere condition, inevitably ensuing upon the progress of molecular events. And precisely and for the same reason that ontogeny abbre-viates microscopic phylogenetic processes, so may microscopic processes be abbreviated. To make this clear, the free animals, A, B, C, D, may develop by gamogenesis or by agamogenesis, *or directly through effect of the enviroment.* A may develop into D. But, beyond this, differentiation sets a bar to this progress. Through the operation of this law low organisms may reproduce lost members, but those highly differentiated are not repro-ducible.

The matter of sexual relativity of cells by differentiation, as well as the hypothesis that an amœboid corpuscle had in the initial stages eaten a ciliated cell, is well illustrated by the fact that the spermatozoon is a ciliated cell and the ovum has amœ-boid likenesses.

The male cell has taken the step of differentiation toward combining the amœboid with the potential planula by assuming the form of a single ciliated cell. The abbreviation or stage of arrest being in maintaining the amœboid monocellular structure with the cilium which belongs to the multicellular planula. The relative female possesses this cilium in potential, *or in the locked up power to develop it* when the proper time comes.

Ova, as we see them, resemble each other morphologically, but not chemically. Protoplasm may be the same, but every egg differs from every other egg in possessing potentials which may be in isomerism, but have not been proven, thus far, not to reside in the protoplasm having *plus* chemical possibilities, differing not only through isomeric arrangement of molecules, but in the intimate chemical composition of the ovum itself. The food yolks, when they exist, differ, and in these there are stored up, ready for use, the molecules to be adopted by the developing embryo in the phylogenetic order of their acquisition by the parent. One chemical combination being made, and structure resulting from it, another chemical combination is inevitable, and the structure must be modified in accordance with this very simple law. The apparent complexity of the process gives way when we study the lower life processes. To the human mind, one and one making two are simple enough, but multiplication, division and subtraction of other numbers always was, and always will be, abstract processes; even after the principles upon which they depend are clearly understood. The disentanglement of simultaneously acting multitudinous forces into their fundamentals, does not impart the ability to predict everything that will occur at separate instants. The mind is only capable of entertaining one thought at a time, and while following a thread cannot duly consider all the other threads in nature. So the intellect contents itself with ejaculating: Wonderful! Inscrutable! Incomprehensible! The integration of the complex, as we see it, follows from our seeing results only, and being compelled to reason out causes.

The ontological copying of the ancestral development is too evident to mention in such processes as segmentation, invagination, etc.

The atomic falling into place may be likened to grains of sand and other substances with different facets, under the operation of gravity, or other force, building up identically two morphological structures where the conditions are precisely the same in both cases. But, as there is a difference between the environment of the amœba and that of the amœboid corpuscle, that difference will operate to modify the structure, and we know that natural selection, phylogenetically, modifies the shape, and that ontogenetic forces are at work to do the same. These forces being in the environment of the amœboid corpuscle and its offspring, constitute natural selection, as do the phylogenetic forces.

The adjustment to the environment is such upon both causes and effects as to compel the modifications of both the reproduced and the reproducing organisms. A process such as reproduction being an inevitable consequence of growth, as soon as it begins will modify the parent processes, and these will react to modify the product, and both together will establish, not only lines of least resistance for their mutual workings, but by the law of development will, whenever necessary, develop secondary adjuvant processes from the environment, whenever a means can be seized upon which facilitates the process. For example, the embryo in the amniotic fluid is under conditions simulating those of its piscine ancestry, and the fluid presence is encouraged because it is adapted to the ontological copying of the phylogenetic method of tissue making; but how does the amnion arise which retains the fluid? The origin and development of the amnion, allantois and placenta can be generally accounted for as structures which ontogeny developed out of its environment to facilitate gestation. The cause of such development of subsidiary apparatus being in the gradual improving tendency of adapting means to ends through natural selection. The chorionic villi, for instance, which answers one purpose, are thrown out generally as are the cilia of the planula, and like those cilia, which serve another purpose, wherever a use for them *happens*, there villi and cilia, both serving the same ultimate purpose, nutrition, develop, or are perpetuated, and, where no use is found, there both villi and cilia disappear. This is precisely what we know does occur. That these subsidiary apparatuses should be deciduous is not to be wondered at, when, after birth, there can be no use for them. If there were, they would be perpetuated, and where, through favorable circumstances in the environment, there exists that tendency, then this gives rise to a form of monstrosity.

At birth the balance is struck to obliterate purely ontogenetic apparatus, such as the allantois, fœtal vessels, etc., by atrophy, through non use, or conversion into some other useful organ.

Extending this evident principle from the embryo backward, it is understood that *development of a developmental process* occurs in all its stages. So that:

Phylogenetic stages will be abbreviated ontologically;

A subsequently acquired phylogenetic structure may, through its importance, arise first;

Ontogenetic organs and peculiarities of development will arise which have no analogues in phylogeny.

And the entire developmental methods of ontogenesis will,

perforce, so adapt means to ends as to make short cuts, and this can and does progress to such an extent through obliteration of unnecessary means, and the adoption of new accessory means, until the most bewildering differences of methods arise by which ontogenesis may copy phylogenesis more or less faithfully, and the means of attaining the end are through simplification, *radically* changed from the phylogenetic.

This may be likened to the means by which the modern mechanic would set to work to construct a steam engine. The original form, and many of the evolutionary steps of the engine, are essentially adhered to, but the mechanic has developed his methods of putting together his better engine and omits those parts which are of no use, or which were of use formerly but are not now, and often builds an important part of the mechanism in advance of the subsidiary apparatus, even though the latter belonged to the older and the former to the newer machine. Then the scaffolding, which represents the fœtal appendages, is, being useless, removed, and though all the evolutionary methods have not been resorted to, the more perfect machine is elaborated and is all the better for the omission of unnecessary stages. But *all* the useless stages are not omitted in embryology, as is evident in our retention of the aural, plantaris and psoas-parvus muscles. The latter, of use only in quadrupeds, is present, according to M. Thiele, in one out of twenty human subjects.

CHAPTER V.

DEVELOPMENT.

It has often occurred to me that eventually, with the brushing away of teleological cobwebs and the introduction of physical and mathematical methods into physiology, there will arise a possibility of computing the lengths of the ages, in years, required to evolve special forms, through knowing the rates of development and involution of organs, and the conditions under which changes occur.

For example: Ancient Greek statuary represented feeble development of the gastrocnemii and soleus with thoracic sizes upon which the rectus abdominis pull had not acted very long. This, with multitudes of other facts, will afford approximate, which may be developed into exact, means for calculating the length of time, under given circumstances, required to develop organs and organisms. And accessory aid could be rendered by geology and archæology, with other sciences.

Taking up the genitalia, the law of differentiation of the general excretory into oviducts and urinary passages, which empty into the common cloaca of monotremes, was followed by the development of a male intromittent organ through impact induration of male ani, and the growth of a sperm canal and final complete external genitalia. The application of which to the complementary organs sufficed to develop the lower canal for ovulation into separate parts for parturition, etc., from the cloaca, which now became anal.

As to the so-called polar forces, which trouble the monicists to account for, in the matter of limb development we know that an organ not used will atrophy.   We often find, among invertebrates, limbs attached to each somite, but with a special tendency to develop those most used and to a retrograde met- amorphosis of those least used, as in lobster development of forceps and telson, and diminution, or at least failure, to de- velop the intermediate swimmerets.   This " fore and aft " ten- dency of effort certainly differentiated the head end through progression in one direction, causing differences in environ- ment contact, as between head and tail.   The natural selection tendency in invertebrata to abort useless organs, ontogenesis would develop; and, inasmuch as a four-legged stool is even better than one with many more legs, as allowing better center of gravity adaptation, the dipnoi dropped all the extra fin ac- cessories of the fish, as the batrachian dropped its tail; the four appendages developed into organs of locomotion, and, as might be expected where, as in the kangaroo or other power- ful tailed animals, the three-legged stool simile served better than the four-legged, then the extra legs are in the way for locomotory purposes and diminish to such an extent that it is doubtful if the kangaroo should be called quadrupedal.

The laws which developed the polar members are those which indurated the tissue of the moving primitive animals by impact.   Where impact occurred oftenest, there development would occur through natural selection, with the secondary elevation of more useful members and extinction of the use- less.

While the teleologist can see only pessimistic consequences from the operation of such a law in nature, to my mind the release from the governance through absence of all law in the fanciful capriciousness of a *primum mobile* which has no analogy in nature, allows the monicist—even the pantheistic kind of monicist—to derive optimistic deductions from the operation of his laws. No matter how the necessity for what develops into a beneficial act or organ may have accidentally arisen, as in the case just cited, continuance of the working of that law in the adjustment of means to ends, continuance of effort tending to benefit either the organism as an individual or society, inevitably will result in the triumph of that which is good for the animal, family, tribe, nation and entire animate earth over that which is injurious. As the world grows older it grows better, from the civilized standpoint (a proviso which does away with the necessity of explaining what is meant by "better"). Superstition, ignorance and crime gradually fade away in the advance of knowledge of the inviolable laws which govern the universe, and the egoistic realization that direct and indirect advantages flow from decency and uprightness directly cause actions to be adjusted in accordance with such knowledge.

We have in the *Pseudonematon nervosum* of Hubrecht the best representation of the simplest quantitative plexiform development of the nervous system. The hydra may show the beginning in its neuro-muscular cell, and I ascribe the failure by others to verify Kleinenberg's discovery to the fact that in a form initiating a differentiation, many of the species are likely not to possess the new structure, even most of them.

This ectodermally excreted compound of phosphorus with ordinary organic hydro-carbonaceous materials brought about a new experience for the cell union. Whenever this compound was disturbed by motions it exploded, providing oxygen had been previously furnished to it. However faint this explosion may have been, it became a matter to be reckoned upon in the cell environment. Any increase of heat in the medium in which the colony of cells lived caused the nerve substance to act more energetically; the "kick" of the molecule was accelerated.

This is true, within limits, of all chemical combinations, and no less in animal tissue.

Light had an effect upon it similar to that induced by heat; it was feebly subjected to actinic influences. Electricity would cause it to act more energetically than any other force, but this influence was only feebly present in the environment, and seldom encountered in the way of shocks. All animal protoplasm is affected by this force; the amœba will be paralyzed into the quiescent globular form when it is directed through it. Mechanical motions within restricted limits caused it to act energetically, provided such molar motion induced molecular changes in the cell contents. Any such latter motion would affect the primitive nerve substance.

It follows that with this state of things, at first doubtless annoying, a readjustment must occur in the colony on the basis of a new mode of life working. The cells then were shocked by the new tissue, but such forms as could not get rid of it through agglutination to the ectoderm, became reconciled to it, through the reactions it caused being arranged against, in its affecting adjacent cell tissue toward obtuseness to it. The highly developed neurilemma encystment is of the same nature as that which covers foreign substances in the body, and is a conserving operation due to reaction of two differently acting substances upon each other. The intercellular distribution of these nerve granules would now exert no effect upon the cells; but whenever by occasional exposure of the granules to an influence which would start the explosion it was discovered that, instead of having to wait for motions to be transferred from cell to cell before the entire organism could be affected by motory causes, this new tissue conveyed the needed stimulus promptly to a distant cell in a very simple way. The next step was to develop an arrangement of these granules in rows, which we see are at first rather indefinite, crossing and recrossing each other confusingly. This arrangement, however, is in obedience to the law which, by final extension, arranges the entire fully developed nervous system in all animals —the law of least resistances. The granules would, from being diffused, be arranged, by the motions of the low animal, in some kinds of lines, even though badly defined ones; the animal must now adjust his motions to accommodate these rapid transfers of stimuli from one cell to another. The quick conveyance of the impressions caused the cells to act energetically, and, in rapid succession, simultaneously; but not always sanely, not always with reference to an object.

Such forms as were overwhelmed by this to such an extent as to interfere with food procuration, and hence furtherance of life, would be killed off. If, by a certain fortuitous arrangement, such as could be effected by the animal motions, the granules fell into lines which would influence the cells in such a way as to assist locomotion (and, it is easy to conceive that almost any arrangement would cause it to move in some way), then the facilities for travel which such cell organisms, such as worms, would thus possess, would enable that form to multiply more rapidly and to escape trouble into which less fortunate forms would fall.

The longitudinal fibers of a worm being pulled together, contracted, the circular fibers would be distended, which would favor the inrush of nutrient material, one of which is oxygen; this exciting the muscle substances to contraction, would force the longitudinal to relaxation and allow it to go through the same process. But the absurdity of a springy worm like this would cause nature never to have invented it without introducing other influences to prevent perpetual motion or establishment of equilibrum with no motion at all. These other influences are the ones operative in all cell movements, with the addition of the nerve apparatus to convey the stimulus to distant cells or to all of them.   How could such a nervous system act to regulate these motions, simple as they are, toward subserving respectable locomotory purposes?

The primitive influence of external forces to excite these motions has not been lost, but there are many internally operating forces at work to change the resultant force, and the worm has an internal environment acting on its cells.   When this resultant is strong enough, through composition of forces, to affect any part of the body more than it does any other part, then mass motion of the colony must radiate from that part. This part, for the time, dominates the rest.   Hunger being the most potent, locomotion will develop in the direction of hunger appeasing motions.   The part which is most affected, the intestinal tract, and that portion of it most in need of food, becomes, for the time being, the center of stimuli production. This influence is telegraphed to the other cells through the nervous system, and various motions in these cells are thus excited.

Least resistant directions influence the motions, and, were one point *always* the exciter, the springy motion would continue till exhaustion overtook the animal.   But the radiating point for excitement must vary, and with this there will be halts, differences of direction, sinuous motions produced, which, with exhaustion or satiety, for the time being, cease; to be renewed as soon as a new stimulus arises, whether it be hunger or some other feeling.   Differentiation decides from which cells and under what circumstances locomotory influences shall radiate.   Development of the function proceeds with development of adjustment to this change, and finally differences in stimuli susceptibilities will arise between the cells. One set will be affected under conditions which will not affect others. Association now steps in to determine which periphery will be linked to a central method of action, or which condition existing in the same periphery will affect the center. The initial start will be made by exertion of such muscles as are relaxed, but action is only possible in such muscle cells as are supplied with oxygen, and where this had a tendency to accumulate, greater action would develop facilities for its accumulation there. The blood, no matter how poorly developed, would, so long as it carried oxygen, be brought to such places in excess of others, and with the development of oxygen carriers, as hæmoglobin, a circulation would result. The general cell need of oxygen establishes an uniform method of circulation The hunger stimulus passes in the line of least resistance; a muscle, the one most often first used, wherever that may be located (usually in the head or tail), contracts; secondary causes radiate another influence to contract other muscles, and where the sum of the influences are met by opposing influences, such as might arise externally or through internal exhaustion, then changes in the nerve and muscle currents must ensue, varying between great activity to quiescence.

When the resultant diffused stimulus arises to a pitch necessary to produce one muscle contraction, if this be frequent, lines of least resistance will be built up between stimulating point and stimulated muscle. Such influences as most habitually follow the first will also build up lines, and the second effect will follow the second cause as the first effect does the first cause, and this mode of motion, by development, will tend to cohere and differentiate other modes. But the natural mode of action may be altered temporarily by operation of other causes. These, too, if frequent enough, are adjusted to and

6

the number of motions possible become limited only by the number of combinations of muscle contraction possible.

In the fish motions the compound curve may be resolved into two simple ones. In each of these one set of muscles has been stimulated to contraction through resultant energy acting in it. The continuance of the stimulus would be meaningless if it did not find another, fresher, muscle to act upon, and the opposite side finding this, the two motions are made; four are as easily made and a general adjustment ensues to make these motions instinctive, but alterable through second and third causes.

A single irritation produces in a muscle, though the nerve, a single "clonic" contraction. The muscle cells develop a tendency, as the conditions become more favorable, to respond *longer* to a given impression. This is shown in some low muscles, requiring thirty irritations per second to fuse the impressions into a tonic contraction, and in the development of the ability in some muscles to remain tetanized with only ten stimulations per second. Helmholz calculates nineteen and one half as the average "voluntary" muscle beat tone.

A single impulse preceding only a single contraction, another impulse must follow the first quickly, or the muscle will relax. It is *easier* to send this impulse in a different direction, to a different muscle, one that is under conditions favorable to its being contracted, and the next impulse will either be sent to the same or another muscle erratically through diffusion, or definitely through definite channel facilities having been organized. When the adjustment is such as to build up channels into inevitable modes of working as regards initial impulse, inevitable sequences follow, and these are taken hold of and develop. One motion made, the next is inevitable, and the greater this development the more *instinctively* is the motion made.

The nervous system adapts itself to the instituted order of things as does the skeleton and other structures. The stage prior to full adaptation of the skeleton is well shown in Cope's fossil batrachia,* where the vertebræ merely held out promises of what vertebræ should be.

Up to this point we have considered muscles as not being separated from the sensitive molecularly-irritable protoplasm. The two properties of contraction and irritability developing in the same cell do not afford full scope for differential motion. The first chemical motions made by the irritable primordial cell must be shared in the same cell by a motion which is similar, but grosser, in its results. That is, the irritability consists in greater molecular mobility, a quick response to stimuli.

Differentiation of these properties will separate off ganglionic gray areas from muscle areas, but it must be remembered that I do not follow the traditional method of calling nerve cells exalted sensory organs. The irritability (facile molecular transfer of force) develops in these separated off protoplasmic cells, and in obedience to the law of development, the cell wall being an impediment to prompt transfer of molecular energy waves, is aborted, and the basis sensitive substance fuses, with here and there evidences of maintenance of the limitary cellular ectoderm and the retention of the cell existence and shape in isolated areas, to be mentioned further on.

I want no mistake as to my meaning here, and hence will deal in concrete examples: The neuroglia of the spinal cord and brain consists originally of a gelatinous protoplasmic substance in which but few if any cells of any kind are visible, the ganglionic substance has run together like so much jelly. This

* American Naturalist, 1883.

is a developed condition. When, as in this instance, it was better that no cell walls interposed between ganglionic areas proliferation would proceed so rapidly as to abort the cell wall, and not until a later type of organization set in, when there came stability of area irritation, do we find cells with walls in this substance.

Differentiation does not extinguish original attributes in cells. Every cell must eat, but where the elaboration of food is carried on by a cell for the benefit of the other cells, accidentally altruistic function through an egoistic process, that enteric or glandular cell must preserve a due regard to maintaining some of its primitive protoplasmic attributes in other directions than eating. If it do not, then fatty degeneration occurs and the fat cell results which may serve as food for other cells in time of dearth. This primitive irritability may, as above stated, develop into isolated cells; and differentiation may occur to various extents. Such irritable isolated cells are always peripheral end organ cells, and by seizing upon accidental materials in their environment may exalt their irritabilities as the visual purple or rhodopsin chemical substance adds to the optic function.

The aggregation of masses of these cells and their fusion, with loss of cell envelopes, constitute the rudimentary spinal cord of the amphioxus. The scattered nuclei, more evident in the marsipobranchial cord than in the pharyngobranchial, are evidences of the cellular origin of the neuroglia, but the quantitative increase of the substance holds in check the other cell characteristics as membranes and nuclei, and, in fact, proliferation from the few original cells of this irritable substance would constitute the ontogenetic short-cut toward accomplishing the formation.

It is evident, even in the spinal cord of man, that segmental areas were the first results of separation. There are intervals in the cord corresponding somewhat to vertebral levels, where cells are alternately many, and then in other levels few. The regularity of this points to segmental fusion.

Figures 11, 12 and 13 represent the stages by which at first

   irritable cells caused direct stimulation of muscle cells. The separation of the two forms of matter and their

FIG. 11.   FIG. 12.   FIG. 13.   communication through plexiform nerve fibers. Lastly, by arrangements of these fibers in lines of least resistance the direct stimulation of muscle is obtained through them, from the neuroglia.

Fig. 14 represents the plexiform fibers which convey in an

 indifferent manner irritations to the neuroglia. Fig. 15 the definition of the sensory nerves.

Could we always depend upon ontogenetic findings as setting forth phylogenetic development, Fig. 16 would illus-

FIG. 14.   FIG. 15.

trate a step by which this develoment occurred, and Fig. 17 the next step, 18 and 19 the others.

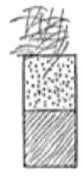 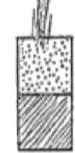 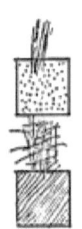 

FIG. 16.   FIG. 17.   FIG. 18.   FIG. 19.

But we know that from the plexus fibers both systems of nerves arose, and ontogenesis often inverts the order of tissue and organ genesis, as in the archinephritic duct appearing before the primitive kidney arises.

Fig. 20 shows the neuroglia segments related by the plexiform fibers, and this explains an embryological feature otherwise inexplicable, thus stated by Balfour:* "The whole of

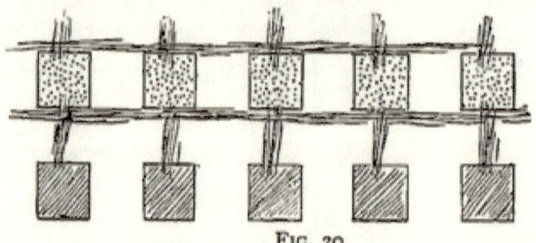

FIG. 20.

the nerves in question" (spinal nerves) "arise as outgrowths of a median ridge of cells, which makes its appearance in the dorsal side of the spinal cord. This ridge has been called by

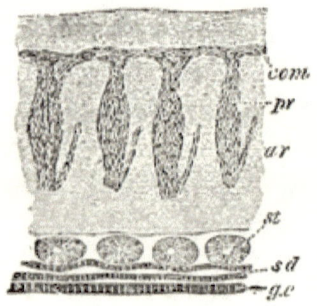

FIG. 21.

Vertical Longitudinal section through part of the trunk of a young Scyllium embryo. *Com*, Commissure uniting the dorsal ends of the posterior nerve roots; *pr*, ganglia of posterior roots; *ar*, anterior roots; *st*, segmental tubes; *sd*, segmental duct; *ge*, epithelium lining the body cavity in the region of the future germinal ridge. From Balfour's Embryology.

Marshall the 'neural crest.' At each point where a pair of nerves will be formed, two pear-shaped outgrowths project from it, one on each side, and apply themselves closely to the walls of the spinal cord. These outgrowths are the rudiments of the posterior nerves. While still remaining attached to the dorsal summit of the neural cord they grow to a considerable size.

"The attachment to the dorsal summit is not permanent. But before describing the further fate of the nerve rudiments it is necessary to say a few words as to the neural crest. At the period when the nerves have begun to shift their attachment to the spinal cord there makes its appearance, in elasmobranchii, a longitudinal commissure, connecting the dorsal ends of all the

* Comp. Embryology, vol. ii, p. 369, et seq.

spinal nerves, as well as those of the vagus and glosso-pharyngeal nerves. This commissure has as yet only been found in a complete form in elasmobranchii, but it is, never-theless, to be regarded as a very important morphological structure.

"It is probable, though the point has not yet been made out, that this commissure is derived from the neural crest, which appears therefore to separate into two cords, one connected with each set of dorsal roots."

He concludes that the commissure gradually atrophies and ultimately vanishes.

This I regard as the embryological evidence of the pre-existence of the plexiform fibers and the derivation of the spinal nerves from them.

If we can accept the illustration and description of the am-phioxus longitudinal nerves in Owen* as exact, the pneumo-gastric and lateral column of the cord formation would be easily

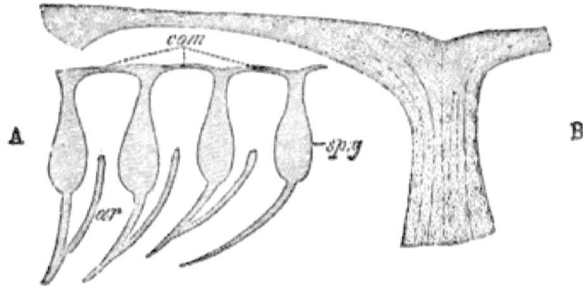

FIG. 22.

Spinal Nerves of Scyllium in longitudinal section, to show the Commissure connecting them. *A*, Section through a series of nerves. *B*, Highly magnified view of the dorsal part of a single nerve, and of the Commissure connected with it. *Com*, Commissure; *sp. g*, ganglion of posterior root; *ar*, anterior root. From Balfour.

explained. The sympathetic nervous system rises above mar-sipobranchii. In this form the pneumogastric is distributed

---

* Comp. Anat. and Physiol. of Vertebr., vol. i, p. 270

to the intestine.   Hunger being the primitive desire, and operating mainly with and against other forces, and tending to render accessory to itself all materials and forces that natural selection had thrown in its way, it seems reasonable that impressions from the intestines should dominate the motions through the early plexiform system.   It also seems reasonable that as the head of the animal differentiated through variability of the environment meeting that end more than the other parts, that hunger can only control the locomotion through being associated with the impressions received elsewhere and gaining control of locomotion.   The great number of impressions received at the head end must act to interfere with the hunger motions, in which case the animal would starve, or it must act to assist them.   Only such forms would survive as allowed the head impressions to cohere with the intestinal for the benefit of the animal. This should be made clear: Grant a rudimentary olfactory sense to the marsipobranchial nasal pit.   Certain molecular rotations of food substances caused a sensation of smell in the animal.   If no notice were taken of this, no association of it with the hunger sense would occur.   If, as is likely, the smell and the hunger attraction occurred at the same time, at first faintly, finally, through persistence, very strongly, came the chemical, attraction, hunger, to cohere with the chemical molecular rotation, constituting odor.

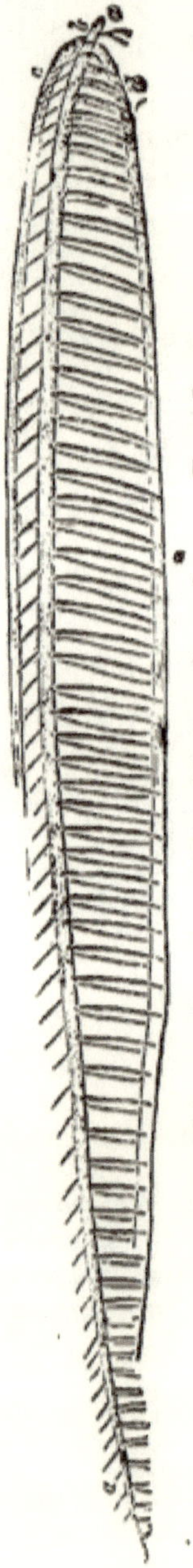

FIG. 23.   The nervous system of Branchiostoma lanceolatum.   From Owen.

Association was inevitable. Where the association was strongest the facility for food acquisition would be developed. Natural selection would pass this new experience, smell, into an adjuvant of the hunger attraction. The adjustment of the cells would be such, when pushed against certain odorous substances continually by the hunger attraction, to make odorous rotations coincide in their manner of affecting the motions with the hunger molecular motions. This coincidence is association. The same power is at work to adapt means to ends in the highest organism, and may be coarsely paralleled by the behavior of many inorganic substances, which adjust themselves to new secondary conditions to such an extent as to often make the secondary conditions the mainly operative. When odorous food substances act upon a terminal sensitive cell (ciliated) in such a way as to start an impulse along the olfactory nerve to the neuroglia cells, and this segment of neuroglia formation is excited into corresponding activity, the activity of that segment occurring simultaneously with the intestinal hunger sensation, no matter where the latter radiated its influence in the production of motions, the two impressions would mechanically build up least resistant lines between them, as may be seen by forming the sand figures of Chladni by drawing violin bows across opposite edges of the plate of glass upon which the sand is strewn. A resultant series of molecular methods of vibrating would inevitably be constructed in that head end segment where the olfactory and hunger impressions met, to excite the motor nerves.

Association, then, consists in a mean molecular or vibratile motion. Under the microscope some of the *results* of this molecular disturbance might be apparent if conditions admitted of such inspection. A mean between the waves sent from two places to a common center. These vibrations may not differ in kind from all others to which the neuroglia is susceptible, except in intensity; the location, the seat of association, being in the higher segment close to the olfactory tract.

This favors the idea afforded us by ontogeny, that the sensory cerebro-spinal nerves arise before the motor.

At this juncture the general spinal motor nerves are performing their force conveyance office to the muscles to move the body, in obedience to the hunger instinct and reflexes; but with the addition of so important an aid as the smelling faculty the food is more directly obtained, and vermicular motions of the intestinal tract are more promptly excited through food contact. The two classes of motions, anterior neuroglia segmental, and intestinal vermicular, acting simultaneously, or almost simultaneously, or successively, will build up association motor tracts between segment and gut, not *for* the purpose of eating, but *because* the eating is associated with that segment irritation. This becomes an advantage. The irritation of the olfactory segment neuroglia excites prompt motions of the intestines in the simplest and most direct manner.

The motor branches of the pneumogastric are built up out of such subsequently exuded particles of nerve substance as the enteric cells may pass into the body channels and intercellular spaces. Instead of passing to build up other nerves, previously formed, they now arrange themselves in a new set of nerve fibrils through operation of the law of least resistances.

All those organs concerned in the elaboration of food for the body, such as the respiratory, hepatic and general glandular, have the hunger sense developed in excess of other senses. It is for this reason that the sensory pneumogastric fibrils pass to these organs as well as to the intestines, and because cellular motions of the same nature occur in these organs, though differentiated, as those in the intestines, motor tracts of the same associating nerve unite the head segments with these viscera as with the intestines.

Let us here dispose of a little physiological superstition: It was considered a great mystery why an impulse over one nerve should result in saliva secretion, and one over another in lachrymal production, etc. This may be understood by likening the nervous system to a pole. If the pole be used to disturb or stimulate a cuttle fish, in the usual order of things you may have ink as the result. If the same pole, or another like it, be used to stimulate a skunk, you will not get ink as a result. The secretory faculty lies in the cell of the non-nervous structure, and has relations to the nervous system the irascible animals mentioned have to the sticks and stones which strike them.

All these organs act through motions induced in their cells by direct influences upon these cells, and until organic development has reached a stage such as we find in teliost fishes, an association system of sympathetic fibers could not exist, because there was not sufficient stability to the organs. With their full development the same causes of simultaneous, consecutive or associated action, effective in cerebro-spinal nerve formation, would now begin in the viscera. A plexiform system of fibrils would creep over and enmesh the organs, and in the establishment of nutrient reflexes build up connections with the cerebro-spinal nerves. Every motion, whether molecular or molar, which induced waste necessitates repair. The motions which accompany all life processes draw blood to the points of activity, and in proportion to the preponderance in this respect of one part of the body over the other, so will the circulatory channels enlarge toward it, and the reflex systems necessary to hasten action in this direction will appear.

Instead of building the sympathetic association system between the blood vessel and point simultaneously active through each muscular motion, making a direct draught upon blood vessels, and thus associating motions. Instead of running the sympathetic fibres between muscle and vessel, the conserving energy of nature would find a more direct means. In fact, were such an arrangement to occur, the muscle ends would atrophy as soon as the cerebro-spinal nerve connection with the sympathetic had been effected. The motion in cerebro-spinal nerve fibers caused the muscle motion, which in turn caused the blood to flow toward the muscle, or rather facilitated it. Association of the muscle motion with the blood vessel motion is transferred through the sympathetic fibers (which facilitate, or, in a certain sense, cause the vascular action) uniting with the cerebro-spinal nerves, which prompt the muscle motion, thus making a short cut for the nutritive process, by which time is saved, energy conserved, and through which differentiation has proceeded and will proceed. As every vascular pulsation has reverberating secondary pulsations all over the body, the molecular nerve motion passes over the vessels to the smallest arteries, and as every spinal nerve action produces wide-spread general vascular pulsations, proportioned directly to size of nerve and importance of function, a commissural line of bands creep along the anterior vertebral bodies and unite all the spinal nerves and the cerebral to the spheno-palatine ganglion through commissures to the vidian end of the commissure.

The great splanchnic nerve is the intermediary between visceral motions and those locomotory and other hunger appeasing motions which operate together

Lack of oxygen, excess of carbonic acid, or other causes, produce vermicular motions of the intestines, which are mainly through contractions of the circular fibers passing downwards. Of course apnœa is induced by too much oxygen and the motions stop through surfeit. These motions are wholly independent of the cerebro-spinal system, though the pneumogastric cognizes them and may incite them to greater activity. Stimulation of the vagus tract increases the motion, while stimulation of the splanchnic, which is of the nature of the fibers passing over the viscera, can only interfere with the low grade motions of these nerves and paralyze the involuntary muscle cells, through tetanizing them, as the amœba assumes the globular quiescent form when electrified. Surgical shock is of this nature.

The splanchnic is, as are other special sympathetic bands, built up through persistence of force and least resistance to accommodate supply of blood to general motor waste of tissue. The vascular condition of large areas is kept in tone through sympathetic ganglia areas and where a distant vascular tonus is changed arterial dilatation, for example, is caused by need of repair in adjacent muscular action. The associated recurrence builds up more direct fasciculi to cause simultaneous compensatory action. The general bodily loss of arterial tonus would, in distributing equilibrium through the general sympathetic and more directly through the splanchnic, effect its compensation more or less rapidly; but where conveyance to so important an organ as the brain necessitates much prompter regulation, the cervical, carotid and cardiac sympathetic distribute the tonus, as is shown when by injury to the cervical fibers or ganglion the head is suffused with blood. The leptocardiac pulsations show that the muscles of the heart are independent of the nervous system for motility. The sympathetic attempts to distribute the pulsations along the vessels in an equilibrating way, but through its cords and gangliar connections it acquaints the cerebro-spinal nerves with the status of the cardiac and vessel tone, and the complicated inhibitory and acceleratory effects are developed by association.

The vesical and uterine sympathetic nerves coming from the same plexus as the colic, indicate that the general excretory and generative excretory functions are in many respects still identical.

Rythmical uterine contractions are of this nature, and react upon the lumbar cord for acceleratory stimulation. Emotional causes may stop the spinal action. Micturition is an equivalent process and subject to the same rules as other visceral movements.

The motor spinal nerve elements are noticeably larger in diameter than those of the corresponding sensory nerves, and this law of sizes holds good for the axis cylinders and associated nerve cells in many areas.

With the conveyance inward to the cord of a multitude of vibrations representing or evoking sensation, the seat of which is in the cerebro-spinal axis, the tendency of adjustment is such as to resist many impressions received in the interests of the organism. It would not answer to have *every* impact and cell disturbance felt so keenly as to cause reflex action. It is in higher life the unexpected impression which does this. In lower forms, as in the sea-squirt, the vast majority of impressions act to project its water. The brainless frog moves its limb when touched, but remains indifferent to other impressions which do not suffice to stimulate the simple or complex reflexes adjusted through past experience.

The evolution of this adjustment is best followed up from the amœba.

This representative of the living cell has its environment adaptation through reaction of external causes upon internal structure to the end that hunger appeasing locomotion may not be interfered with. The compromise is made with the surroundings. At first, any change from the normal would cause a motion which is equivalent, for our purpose, to a sensation, and identical with it. The recurrence of these changes must be adjusted to by molecular reintegration. If the changes act in the food-procuring direction, the molecular arrangement is integrated with the hunger arrangement and associated action occurs. The operation of the change, then, is to excite and promote hunger motions; but if the changes oppose food-procuring, the animal must die unless a compromise be made, and it depends upon the ability of the molecules to readjust themselves against the change whether survival occurs or not. In many it does not—may be in most of the primitive animals. In some it does, whether accidentally or not, the adjustment will be transmitted to the offspring and perpetuated. This may be through induration to blunt reaction, or through softening to disperse it. The protoplasm may be likened to a strong attractive energy seated in a mobile mass of granules which, however the environment may operate, often, not always, in time, through the resultant changes wrought, enables the native attractive ability to be exerted in a modified way. Those less modified are perpetuated and multiply, but inevitably a re-arrangement is effected and new relations with the outer world are established. The injurious impression ceases to become a stimulus; the favorable one is taken advantage of, and continues to excite reflexes until finally it unconsciously does so; the molecules being adjusted in an instinctive way to oppose as little resistance as possible.

So those impressions which cause quivers in the central

neuroglia always affect spinal motor nerves according to segments excited. If a sensation does not produce the quiver
(cognized by higher animals as consciousness), then the sensory impression must have been adjusted against by resistance
arrangement of nerve granules, or, in some other way, something is interposed to minimize or destroy the impression, as
one. Foot callosities are the grossest method of accomplishing this, and here inner to outer relations are compromised.

The same result could be attained by lessening the nerve
fibrils in number or sizes of cross sections to a part, or by
atrophy of sensitive peripheral cells. What physical law is
there that introduces counterbalancing agencies toward conservation of living structures in such instances? With man,
injurious surroundings excite attempts to escape from them resembling the indefinite movements made by low grade animals
in their reflex movement to all impressions. Failing in this,
man makes the best of the matter, and in time may thrive under the change through finding some elements in it which may
be turned to account. Disagreeable, *i. e.*, hunger intensifying
impressions,—ultimately resolvable into such,—promote greater
cell activity if mass motion, which is first evoked, does not
suffice to remove the cause, then the continued cause acting
upon large areas of cells make extensive changes, until a sufficient equilibrating change is effected somewhere, and a balance is struck. The hunger appeasing motions have now the
victory and the obnoxious impression has ceased, even though
the cause still exists. The impression may be disposed of in
many ways: diffusion channels may distribute it over a large
surface of sensitive tissue by development, injurious impacts
may disease (reintegrate) nerve or other tissue destructively,
so as to resist the impression in varying degrees, or the epider-
mal thickening through stimulation of cell growth may blunt

the effect. In which case the sensory strands tend, through non-use, to lessen their sizes and numbers. So, while there are a few causes, as in the optic and auditory sense differentiation, which tend to develop sensory nerve fibrils in certain areas with conservative natural causes, keeping them and their associated apparatus within useful development bounds, there are many causes incessantly at work to make such development undesirable, and retrograde changes incessantly operate to increase resistance of sensory nerve fibrils by lessening their axis cylinder cross section, and in other ways, resulting in evolution of useful and destruction of useless or injurious sense channels.

But as the adjustment of the inner to the outer workings is through interposing changes between the cerebro-spinal axis and the periphery, motions of that axis once set up radiate over motor nerves which are not subject to the injurious influence of an often inimical world. The inner tissues are on adjusted friendly terms with each other. They have shaken down to smooth working through egoistic necessity, which always, when intelligence progresses, finds that mutual assistance is beneficial to self in the highest degree. The frictionless workings of the motor apparatus have been acquired through ages of inherited adjustments, and hence motor nerve fibrils are adapted to their work, and seldom being retrograded, maintain a relatively larger size than the sensory nerve fibrils, for the reason that where in the latter tissue resistance is a desideratum, the minimization of resistance through increased size of fiber is equally desirable in the former.

Rythmic motions of organs, such as blood vessels, and the establishment of cyclical phenomena, such as menstruation, seem to me to have arisen thus: With the constant presence of oxygen in the air and water, the entrance of oxygen causes the amœba to move. The molecular rotations unavoidable in chemical interchange would expand the albuminous tissue and the water presence in the mass admits of this rotation. The disengagement of carbonic acid which follows, is another source of motion. These two motions, unlike the general assimilatory, would act toward increasing the bulk of the animal, if both gases were retained. The carbonic acid being excrementitious must be thrown off. It exhales from the ectosarc or accumulates in the vacuoles with the water there. The oxygen expansion acts with the food expansion to finally propel the vacuole contents outward. The animal would burst otherwise. The passage outward of the vacuole is not always seen. It may, by the growth pressure, be broken up into minute particles which pass between the granules and through the ectosarc. This matter must be projected outward or the animal would grow indefinitely or burst. The latter event would liberate the gas if it came to that pass, and the exhalent process is like a disruption. Whatever may be said of the other pabulum, oxygen is generally present, and CO_2 exhalation must occur after the oxygen inspiration. The one act follows the other, for one being the cause of the other the act of expiration can but follow that of inspiration. This constant action produces the rythm observable in the amœbic vacuole contractions, and where that rythm is absent it is through interference to which all animals are subjected, and irregular pulsations or respirations follow.

In the intestines this downward peristaltic motion is due to oxygen and tactile impressions, and in the highly developed lung the oxygen and CO_2 exchange cause the same rythmic respirations as in the amœba. In the arteries and heart the waves of pulsation occur as the hæmoglobin of the corpuscles yields up a nutrient modicum to each cell as they pass. These cell movements stimulate reverberatory molecular motions in the sympathetic, and these fibers convey their motions feebly, stimulating the muscle cells ahead to contraction and the tonus of all the vessels are thus brought about.

If, through lines of least resistance a special strand of the sympathetic relates distant parts, then the stimulus will skip a large area of intermediate cells and act upon the distant part direct.

The splanchnic thus telegraphs the tonus condition from the intestines and stomach to the general voluntary muscular system and entire body through the commissures on the ventral part of the vertebral column. The cervical and carotid strands relate the vessels of the brain with lower levels of vessels, etc. When the ovum has reached a certain stage of maturity it provokes profound changes which must be accounted for. Simultaneous development of several ova are likely to occur where the conditions are the same in the body, and such full development appears to require a lunar month.

At the expiration of this time the excretory desire of the cell provokes expulsive motions, the growth of the ova produces an irritation on adjacent parts which advances to an extreme. Associated cell action use up more oxygen than usual, and a process equivalent to a local inflammation occurs. More blood is sent where there is increased activity, and through associated habitual action the engorgement becomes useful if the cell be impregnated and thus possess the requisite chemical affinities to exert other effects upon the uterus. If the ovum is not impregnated, it has not changed from its early excrementitious nature in lower animals and is cast out with the released menstrual blood, which at this stage is also useless, setting up a physiological hæmorrhage varying in quantity, duration, painfulness or regularity with different females, showing that full adjustment has not occurred. In fact, the inconvenience and sickness thus produced upon nearly all females show the pathological origin of the function.

The development of gestation from the oviparous to the viviparous is merely through longer retention of the embryo in the mother, and some reptiles and amphibia are known to possess the faculty of breeding in either way. The rule being that higher development is obtained through such longer retention, and applied to sociological matters the analogy holds good. Higher development is obtained for children through fostering of parents than where they are turned loose to shift for themselves too soon.

With the mammary appearance, an associated action of the uterus and ovaries is made. In marsupialia the excitement of the mammary glands by the young follows their birth. At the monotreme origin of this mode of nourishment but feeble effect was produced by the attachment to the breasts; subsequently, where blood was drawn, epithelial induration was caused, which formed the teats, and the epithelial cells thus produced were, with their increase, eaten by the offspring. The blood increased the quantity of these cells to the parts as the parts were drawn upon, and the reparative process is converted into a maternal act of nourishment. These cells, instead of passing to the dermis, were sucked through canaliculi, and in the immature condition for reparation passed with the serum drawn from the blood of the mother to the young animal. The vessels had adjusted against blood, as such, being drawn, finally passing out with the serum only cells in a state of fatty degeneration, constituting the milk globules which, through adaptation, become the easiest yielded up and the most nutritious to the offspring.

With this development came the ontological preparation for

these glands, and with the stages of growth mammæ grew in keeping with the body of the female.

The birth preceding the breast attachment through probably millions of years, the blood vessels associated the two irritations in sequence, so that when the ovaries were excited the breasts would be also.

Ovarian and uterine agitation and growth become associated by simultaneous disturbances with breast growth, and the blood vessels thus simultaneously acted upon form in lines of least resistance to carry on the associated function. The sympathetic nervous system along these vessels ending in glands in a double row from fore-leg to groin, originated just as we described the simultaneous occurrence of odor and food pressure caused the production of the pneumogastric. When the ovary or uterus was then irritated, the breasts would be stimulated to degrees of excitement, as odors produced the hunger motions of the intestine. The converse in both cases is slightly true. Irritations of the breast are sometimes felt in the genitalia, and hunger in primitive animals may produce feeble odor hallucinations. In man, the olfactory sense has diminished to such an extent as to be of but little use, and hence seldom aroused to hallucinations.

Vicarious menstruation through the nipples is thus accounted for by an imperfect adjustment of the process.

During normal menstruation the outflow checks the further disturbance of the breast. During gestation the blood is distributed between breast growth and embryonal nutrition.

The hunger theory accounts for the female desire, but upon what principle can we base the corresponding male desire? In amœba, monads, etc., the relative female or germ cell eating the relative male or sperm cell, on the part of the latter was a somewhat passive transaction. The hunger appeasing of the female was followed by its plethoric quiescence; and in the case of an amœba which had eaten a synamœba segmentation would begin and the amœba released developing into either form or both forms, depending upon the degree of ectosarcal preponderance. At this low stage to eat or be eaten could have no such significance they have in a higher life. It was a process of overcoming. The amœba, with its low ectosarcal interference, could eat the synamœba better than the latter could the former.

Sociologically the money-grubber devours the services of men of brains, and the issue of the business is development of faculties and facilities for mercantile improvement both in the sordid and mental aspects.

The adjustment of the relative male toward perpetuation involved his being finally eaten to enable him to survive at all, and with the rapidity of his new birth and fusion with the female, it is conceivable that in this fusion an indifference to the process would ensue from the chemical aspects as to which ate the other, the only question being one of feasibility. The impediment to the process would be the ectosarc degree. The one which had less of this impediment would be the eater, and the one with more the eaten.

But as the developmental stages advanced and copulatory acts grew from this, it is not so clear how this sexual union could afford any gratification to the absolute male, save through remembering that excretory necessity is transferred from the organism to its cells and the sperm cell increase (capable of stimulation through revivification of causes which before stimulated it) is associated with excretory desire. Through gradual evolutionary associated method of excretion, from fusion of germ and sperm cell direct to hemaphroditic fusion desire between the products of ectodermal and endodermal cells, the gratification of the cells became a mode of gratification of the tissues and entire colony of cells. When sexes developed into separate organisms, the cell desires which had been operative during millions of years of evolution were quickened into existence when, at first, accidental contact (as in juvenility, when the organism first is astounded into a recognition of this, to it, new experience), next through memory and association, step by step the process has grown with development of the genitalia to its present modes of working. The excitement of the entire body in this act is the recognition by the organism cells of the necessity for it, and the realization of a suppressed desire, intense in proportion to its previous deprivation, less intense with satiety, precisely as precedence of pain or pleasure makes all desire purely relative.

The trout and other teliosts recognize the spawn through this association having been connected through the optic sense and the sperm ejaculation is excited. This associability is transferable to such a degree as to account for first one sense then another, may be several senses, as olfaction in dogs, the optic in fishes, and auditory in mosquito male, and the tactile in all, being respectively alternately and together excited. In man, the optic, through association, from savage to civilized experience refers to a slightly exposed ankle what before was only generated by contact or a glimpse of other parts.

The interdependence of the tissues is such that an abnormal condition arising in one point may set up many changes in distant parts. For instance, a change in the assimilative abilities of an intestinal or glandular cell may chemically change what it excretes, and by transfer alter the food of remote and remoter cells until excreted. Itching, pruritus, and other obscurely caused conditions, as urticaria, which is induced often by eating shell fish, have these relationships of cause and effect.

Profound changes were entailed by the exchange of the watery medium of the Sozura, or tailed amphibia, occasionally resorted to by them, and the advance to the protamnia stage.

Not only were the gills aborted by this loss of the watery environment in which they were bathed, and the lung developed, but the amniotic development became necessary and was a consequence of having left the water, whereby ontogeny substituted an accessory means of copying phylogeny, and the embryo was passed in utero through the aquatic stages to the pro-reptilian and pro-mammalian stages.

The organ of tears originated at this time, by the painful dessication of the eye ball front exposed to light, rendering an occasional dip into the water by the amphibian necessary, at first, until adjusted balance of water distribution thereabouts could occur.

The water received by the eye-ball from the sea or river grew less necessary as the outpour from behind the eye became facilitated, and the lachrymal gland was created. This was a purely mechanical process, similar to the bibulous capillary attraction of blotting paper in its distribution of water in one part to the other parts. The dry cells of the anterior part of the globe drew upon the wet cells further back, and as the adjustment to this draught enlarged the water capacity for the rearward cells, a certain group developed into the function of furnishing this fluid to lubricate and protect the eye, and a continuous life out of the water was made possible.

The pain of the light-glare thus associated with the tear-flow afforded an important serviceable associated habit for a multitude of grief and pain expressions which subsequently arose. Previously, pain was otherwise associated.

The dropping of the tail seems throughout to have been a promise of higher cerebral development, and where the caudal vertebræ have persisted in any line, the development of the encephalon has been slow in that line. The tail seems to have persisted in our phylum to the prosimiæ.

In most invertebrates there are no morphological distinctions

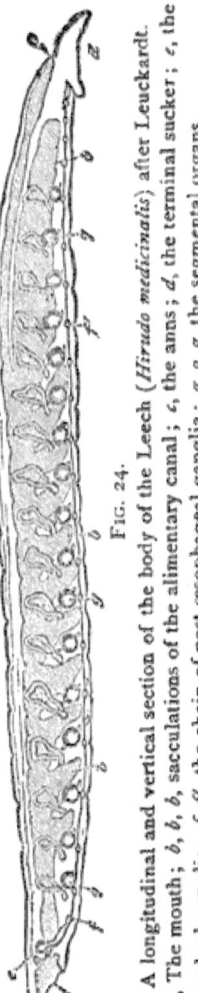

FIG. 24.

A longitudinal and vertical section of the body of the Leech (*Hirudo medicinalis*) after Leuckardt. *a*, The mouth; *b, b, b,* sacculations of the alimentary canal; *c,* the anus; *d,* the terminal sucker; *e,* the cerebral ganglia; *f, f,* the chain of post œsophageal ganglia; *g, g, g,* the segmental organs.

of sensory and motor nerves, and as in low worms we have reason to believe the two functions are not separated, we will have to assume that the muscle cells are mainly stimulated by transference of irritation from one cell to the other; but the ganglia cells, which are originally neuroglia, or, as we may term them, sensory, are united by a commissural cord (*Hirudo medicinalis*). These ganglia correspond with the somites, except where fusion occurs. Leuckart * counts, in the leech, thirty pairs of post-oral ganglia, but the seven posterior and the three anterior pairs coalesce. Nerves are given off to the pharynx and intestines, and the former develop special ganglia. The post-oral commissure we may look upon as a sensory nerve relating the sensory ganglia; the filaments given off from the ganglia may be regarded as motor. All impressions pass from the head and somites to cause a contractile wave.

Directly upon the anterior segment are placed simple eyes, nerves from which pass to the supraœsophageal ganglia.

The feebly appreciated impressions are carried backward

* Huxley's Invertebrate Anatomy, p. 191.

to, and through, or by, the sensory ganglia, thence radiating through motory nerves to somite groups of muscle cells.

In the transference of stimulation of one cell to another, the observed facts of the nervous system afford simple explanations of their modes. The granular fibrillæ lie upon and among the cells in direct contact with them. This is true of sensory gangliar matter as well as all glandular and muscular tissue. When the fibrillæ are naked, of course, large areas of cells are affected through direct impressions; but when they are medullated, isolated, insulated, action carries the stimulus to a distance.

It was very puzzling to histologists to see how simply the nerve fibrils ended, spread over muscles and projected into spinal cord gray. It showed that contact of the nerve fibril only was required to bring out the peculiar mode of action of each cell when excited. Were the hirudinea sensory nerves connected directly with the muscles, then a nonsensical rigor would perpetually seize the animal, without definite end. The interposition of the sensory cell set a bar to this by introducing switch points to muscle areas and affording the muscle nerves a *reliable and constant method and source* of initiating motion. Instead of millions of different points acting erratically upon muscles direct, the myriad impressions were sent to a few ganglia, and these radiated the impressions to the muscles, as from reservoirs, making the motions definite, and imparting regularity where otherwise there would be confusion.

It is only necessary to imagine that this jelly-like neuroglia substance of the ganglion area quivered, and that this was produced by nerve irritation, and in turn produced nerve irritations ending in definite muscular contractions.

With the fusion of ganglia and the constant passage of nerve influences in a certain direction, from head to tail, changes in those structures must occur.

In the acrania but little difference exists between cranial and spinal nerves. None of them possess intervertebral ganglia. Free nuclei abound in the ganglionic masses, and the next step is a quantitative development of the cranial ganglia, owing to the greater frequency of impressions there. Then granules abound about these nuclei, interspersed without much order in the lamprey cord; finally, in higher animals, assuming fibrillar and clustered arrangement. The nuclei afford the centers for this clustering, but the direction in which the neuroglia quivers pass would arrange the fascicular rows and form the columns of the cord, becoming medullated if there were differences in modes of action of the two tissues, as a resultant of the two modes. Such new medullated fibrils as remained uncovered in the neuroglia substance indicate modes allied to those of the matrix from whence they sprung, and their directions indicate the tendency of the neuroglia quivers to pass the tremors in certain ways. Failure to establish thorough uniformity of direction indicating the conflicting nature of the disturbances diffused through the cord. But a resultant arrangement will be compromised upon and the fascicular picture of the fibrillæ show strong tendencies toward arrangement. Most persistent directions and complex arrangements show the presence of contending directions.

From the neuroglia substance by proliferation of eminently sensory cells, of which the scattered nuclei are remnants, and the loss of their cell walls, which was a developmental feature, arose again a disposition to reform the cell wall, when through clustering of granules about a nucleus this was made possible. The granule parturition of some of these embryonal cells is excellently shown in the researches of Schmidt.*   A cell becomes surcharged with granules, the wall bursts, and liberates the granules which mingle with the general fibrillæ. Sensory cells of the neuroglia are seen in the morula stage, and so-called mother cells are seen to divide by fission, as does the amœba.

Association of impressions, or the simultaneous action of two sensations in lines of least resistance, build up the filaments which go to form the sensory strands on the posterior and postero-lateral columns of the cord, and simultaneously acting motions also build up the strands of the antero-lateral and anterior columns.

Every nerve cell contains granule contents.   The shape of the cell and the direction of the granules are determined by the composition of forces.   This cell merely admits of definite transmission through it of the motions common to all nerve fibrillæ, and owing to the comparative fixity of these directions the cell wall binds the fibrillæ in place and affords it a localization.

Herbert Spencer claims that function precedes structure, and in the evolution of mind from chemical affinity (a function of matter) this is evident.

We have somewhat fully considered structure, or what function has integrated; now let us turn to the effect of function upon structure, and show how the mind has reacted upon the body to build up the brain.

In the amœba we saw that chemical affinity acting in its protoplasm constituted hunger, which was the primitive desire, and that locomotion was effected through gratification of this hunger or through the desire to gratify it.

---

* Journal of Nervous and Mental Disease, July, 1877.

The sexual appetite was differentiated from the hunger appetite as soon as special cells developed to carry on the sexual ingestion and excretion.

The tactile sense feebly separated from the hunger sense as soon as cilia developed, and the locomotory tactile as soon as organs of locomotion were fully developed.

The remaining senses arose by differentiation, but up to the time a nervous system formed, the too independent action of these desires and functions must have operated but feebly to conserve the general good of the organism. With the advent of the plexiform nervous system no one cell could suffer pain or pleasure without instant participation therein of all, or most of, the other cells. This became an advantage comparable to the linking together of continents by electric cables, which, apprising distant parts of activities therein, attune the general behavior accordingly; and who doubts that, since cables have been laid, wars have been rendered less possible, and that a better international understanding has been brought about?

Instead of the worm-cells warring with one another, and an impression reaching one part of it long after the possibility of the organism to act in accordance with the impression had passed, now the stimulus was instantaneous, and the cells were brought into close sympathy with each other, where before, through cell to cell influence, this sympathy was feeble and often ineffective. A new resultant from the summation of internal cell-activities arose, but all must be subordinated to the food-acquiring ability which throughout all animal life must dominate or the animal perishes. However differentiation may have masked desire, the fundamental hunger guides the motions, and, having its base in chemical affinities, man is attracted or repelled, as the animals are, from that which is pleasureable or painful. The boor, generally, thinks only of his food or of some of the lower offshoots of the appetite for it, and in the scale of intelligence of animals through to the savage, barbarian, and civilized man, our only knowledge of the intelligence of others is derived from witnessing in them motions better and better co-ordinated toward the food-acquiring and allied functions, and finally the development of such facility for food-acquisition as to make it hold an apparently subordinate place in our reflexes, with consequent leisure and ability to develop and differentiate in directions which, though based upon hunger-appeasing, apparently, to the unthinking, have no reference to it whatever. *All* desire is based upon it, and the developed desires fade away in inverse order of their creation precisely as the fundamental ability to get food is interfered with. Whether starving from sickness or poverty, no artist or scientist can think of anything but his need, and if, through rebellion among the cells of the organism, or through indifference of neighboring animals, this extreme arises, the result is the same in degrading toward the basal instincts and desires. Where a "ruling passion" struggles through adversities to manifest itself, it must be strong indeed, and society has fought all innovators toward beneficent development from the law of cell-resistance to foreign materials until they gain a foothold. Then, society and the cell complacently adjust to the new order of things and plume themselves upon their shrewdness and ability.

To conceive of what the lower animals think, we may resort to the objective method of inspecting their motions as they grow more and more intelligent, in proportion as their cells are internuncially related by the nerve-fibers, and subjectively by remembering the sluggishness of our own motions and sensations when we "feel stupid" or are falling off to or awakening from sleep. Objectively we thus judge of insanity.

In certain diseases of the nervous system the nerve-fibrils do not convey the sensation, and the brain receives the impression slowly through the gray matter of the cord, or until a higher, unaffected strand can pass it upward. Many seconds elapse before a pin, stuck in the body, is felt. This is a pathological dissociation, which is extreme as regards sensation in anæsthesia and as regards motion in paralysis.

If the sensory and motor chemical acts of the amœba are judged as constituting its thought, then to think of something to eat, to feel the desire to eat, and to move with reference to getting food, are one and the same thing. When the cells are so separated as to perform different offices, each thinks mostly of its peculiar function, which, with reference to the associated cells, may constitute a difference; but with the cell itself the fundamental desire has changed but little. Thus the cells devoted to reproduction, hunger, and they desire to excrete, and in so doing are related to the organism as establishing a difference of mode of action from the lower cells.

The act of the muscle cell is an inspiratory eating one when it contracts, and so on throughout the body. The seat of the action constitutes the sensation difference and the quantitative development of cells devoted to the performance of a function determine the intensity of the feeling. The nervous system merely converts the irritation from these seats into vibratory terms and conveys it to the *sensorium commune*, the neuroglia, which is more responsive according to its quantitative development, its aggregation into a cerebro-spinal axis, and its being furnished with association fibers to relate distinct parts. Hence *mind is located in every living cell of the body*, and the better nervous association of these cells constitute grades of intelligence. Cells vary as animals do between themselves. The most molecularly active cells, such as the gangliar basis substance, have developed an irritability beyond what the amœba possessed through differentiation of its molecular mobility, that being the activity which best appeases its hunger.

When, after rest and the restoration of nutrition, we feel well and strong, every cell desires to carry on its function and excites the body to general activity.

Such cells as have developed the highest degree of molecular activity, have, through this development, acquired an exalted mental activity, but while thus the neuroglia is the seat of this exaltation, it is one of degree only, for every other cell cannot be denied mentality as long as chemical interchanges go on in that cell. Consciousness increases with the number of cells properly related, and with the facility of tissues to undergo chemical metamorphosis. Mind develops with the coördination of these chemical relations, a coördination which is possible only through phylogenetic and ontogenetic processes. The pantheistic applications of this to all inorganic nature I have nothing to do with in this essay, and leave deductions of that kind to others.

With every impression received, every cell impressed is readjusted. It assumes a different shape, its molecules and granules are differently arranged, and varying densities are instituted in different parts of the cell, from colloidal conditions to states of rigid immobility, depending upon the cause and the nature of the cell and its contents. With this, adjacent and remote cells are affected, as no cause has but one effect. Nature is full of the operation of this law. Given one cause, the remote effects flowing from it are inconceivable in extent or importance.

The better adjustment of the cell to its work constitutes *instinct;* the hesitancy in response to the cause, through resistance to the new arrangement, may be called *thought;* and whenever the effect of the first cause is reproduced, no matter how, whether by direct repetition of the first cause, or by association, or by any extrinsic influence, this repetition of the first effect constitutes *memory.*

Instinct, reason (thought), memory, are thus but modes of cell adjustment and operation ; and the instinct producing less disturbance than the reason impression, so the memory exercised in reasoning is of a more noticeable character than the memory evoked in instinctive motions, because the instinct and the instinct memory are facile workings of the cell, while reason and the reason or thought memory, are difficult, in degree, workings of cells, in which or for which adjustment is incomplete to the impression. As soon as the adjustment is made the reason, thought, ceases and instinct begins.

The inevitable consequence of this is to hand mind in all its manifestations of reason, instinct, memory, etc., over to the general body cells, with mind *par excellence* seated in the cerebrospinal gray matter, which may be called, though only relatively so, the seat of consciousness.

There is then a motor mind through the motor apparatus adjustment made up of a motor instinct, reason and memory. The latter was considered by Küssmaul * as a necessity from conditions of aphasia, such as inability to pronounce a word even though it was in the sensory mind. He called it " *Bewegungsbilder.*"

Reviewing the processes of development of the central nervous system, its simplicity is wonderful ; and Spitzka† affords us a succinct account of the steps, which may be summarized thus :

Two axial gray ridges rise up on each side of the embryo : these ridges fuse to form a tube around the central ciliated epithelial canal.

---

* Berliner Klin. Woch. 1870, Nos. 37, 38.
†Architecture and Mechanism of the Human Brain.

7

" The first indication of the cerebral formation is a pyriform enlargement of the anterior end of the medullary tube, which, like the parent tube, is hollow. The walls of the enlargement (primitive cerebral vesicle) are composed of the same elements arranged in the same radiatory manner as the walls of the rest of the medullary tube. Through all subsequent morphological changes, the cellular elements of the brain hence maintain a vertical position to the surface contour of their respective locality, unless their position is disturbed by interruptory fibers.

" The primitive cerebral vesicle exhibits two constrictions which subdivide it into three lesser segments, each of which contains a more or less round cavity, connected with the other cavities of the constricted lamina. These vesicles we designate as the (1) *fore brain,* (prosencephalon); (2) *mid brain,* (mesencephalon), and (3) *hind brain* (postencephalon)."

The thalamus and hemispheres constitute a structural continuity and rank with the corpus striatum as such.

" Water breathers constituted the ancestry of the vertebrate sub-kingdom, and it is reasonable to suppose that the olfactory lobe constitutes one of the most ancient differentiated segments of the central nervous system. In the myxinoid fishes and the lamprey the *cerebral hemispheres themselves are mere appendages of the olfactory lobes, and hardly half their size.*"

Morphological considerations are of secondary importance in a psychological essay, but I have never seen good reasons for vacating the ground I took in declaring the intervertebral ganglia of the spinal cord to be homologous with all the tubercles and lobes of the brain and the laminæ of the cerebellum*.

The Gasserian ganglion is indisputably an intervertebral. The olfactory lobe of all animals (in man degenerated to a tract) is an intervertebral, the optic lobes of the teliost fishes are homologous with the intervertebral, and are erroneously

---

*Journal of Nervous and Mental Disease, October, 1880. American Naturalist, January and February, 1881, and July, 1881. Chicago Medical Journal and Examiner, November 1880.

called its cerebrum; the globular shape of the cerebellum lobes, even in highly developed animals, and the passage from these globules to a laminated form in quadrumana and man, and the fact that all such ganglia have only sensory nerves passing through them, are the main points. The reasons in full and extension to aborted lobes, such as the pineal gland, etc., are given in the papers mentioned.

However subsequent ontogenetic fusions may picture the intervertebral and the spinal gray as morphological units, Balfour* distinctly shows these swellings of the posterior roots to be developed in scyllium, independently of the medullary tube, and the nerve root, with its anterior companion, subsequently join the tube.

In its earliest stages above the lancelet such an interposition of primitive sensitive substance could operate in no other way than as an *etape* to blunt the incoming dorsal sensory impressions in the interests of other and cephalic ganglia.

The forward appearance of these ganglia are significant as intermediating certain impressions before allowing the cord to be affected, such cord impressions indicating inevitable motor response.

The olfactory lobe or intervertebral thus quenched certain useless sensory impressions which, before, were mixed up with the pneumogastric hunger senses, and at this stage higher discrimination began.

The question of vertebral origin of the cranial bones seems to have been settled by Huxley's dictum that these latter arose at the same time with the vertebræ, and through partaking of the segmental peculiarities the cephalic modifications render them *sui generis.* So with my intervertebral homology. In

---

*Comparative Embryology, Vol. II, p. 371.

the main, I believe it to be true, but in this essay it does not become an essential beyond best accounting for developmental processes.   It suffices to regard the fact of secondary gray interposition upon sensory nerves, no matter how it arose.

With this suppression of certain vibrations of no immediate use to the animal, a discrimination caused by physical processes eventually came about, a determination through commissural erection of association of these hitherto useless impressions into useful ones.

This should be made clear:

I claim that every impression reaching the spinal cord produces a corresponding muscular or glandular motion somewhere, unless inhibited by a secondary apparatus, to be regarded further on.   Any way, the motor apparatus is affected.

Such as olfactory sensations occurring simultaneously with the hunger sensations cause a tract between the points of reception of these sensations in the gray to be affected, and subsequently linear arrangement of primitive pneumogastric fibrils.

Many mistakes must occur and the excitement be produced when there is no food present and the animal is " fooled."

But with the tendency toward differentiation of impacts these especial vibrations, which are unerringly associated with food presence in the olfactory, continue their former effect, while the vibrations which were mistaken for food odors affecting the ganglion cells (neuroglia) in a different manner, however slightly, became *dissociated* by the most natural means. Thus we see in intervertebral ganglia fibrils passing through the ganglion and others ending in it.

These ganglia *feel* as do other cell aggregates, particularly such sensory cells, but as they are comparatively isolated what they feel does not affect the organism until with commissural union such ganglia as the cephalic unite their modes of motion with others.

Transverse commissures are attracted by simultaneous excitement in kind of bilaterally placed gray masses.

Longitudinal commissures are built up by simultaneous or successive excitement in kind or totally different excitement occurring in different levels of the body.

When the "shunted" motions affect the organism it becomes a conscious sensation, evoking at first no definite motion except perhaps *diffusion*, if profound enough evinced in trembling, etc. In fact, the secondary ganglia apparatus is a diffusive apparatus, and we may so regard it. It is filled with embryonal fibrillæ, which are acted upon in all directions diffusively, but which do not exert any definite effect until through least resistance the plexus forms a tract which serves a useful purpose. As everything in discrimination or judgment depends upon association, it is easily seen how this could result in making many errors. For instance, two simultaneously occurring impressions build up, from the brain plexus of fibrils, a tract which, by associated action, dominates the motions, excites, say, the feeling of fear. Now, while this simultaneity may occur, the association, objectively, may be purely accidental, and the animal will run from an impression which has no harmful reality. Eventually, through differentiation and the greater persistence of the truly harmful impression without the harmless one, a modified tract dissociates the two, and by this process of integration, disintegration and reintegration, all acts of all animals with nervous systems are regulated. Consciousness, growing from the chemical affinities, increases in direct ratio with the masses of molecules acted upon, with the intensity of the chemical reactions of such molecules; but the consciousness of the organism can only become such through the associated commissural union of such chemical molecular changes, and the greater and more definite this union, the greater and more definite is consciousness.

Change, difference through resistance, and the necessity for overcoming, render consciousness more acute; and when, through adjustment of the molecules into the least resistant methods of working, consciousness passes into instinct, reflex motions thus instituted become inevitable ones, and, acting less vehemently upon the sensory cells, do not excite consciousness at all, or but very little.

As psychic and physical life are inseparable, the mind has merely developed as a complex upon the better and still better method of conserving the physical. Many mistakes occur in this endeavor; but, nevertheless, that is the *intent* of the mind. That failures are made is shown by disease and premature death; but the survival of forms shows that they have adapted inner to outer relations more or less effectually.

The senses are but relative forms of motion, and are unclassifiable except as they manifest themselves in groups, such as the common five senses. A sense may be any molecular motion possible to a cell, or at first not possible, but into which it may develop, providing such motion produces profound enough impressions and serves a useful purpose by association. In forms in our phylum the olfactory was at first most important, but other senses have superseded it, and "shunting" is progressing at such a rate that this originally peculiar tactile ciliated cell-motion has become very obtuse in man, and few fibrils project through the olfactory nerves to unite with other strands, and such as have originally done so are being diverted in the formation of more active sense fasciculi. This development and destruction of sense is obvious everywhere among men who become obtuse to certain impressions and develop extraordinary susceptibility to others. The auditory, another differentiated ciliary tactile, has developed to an important position in life conservation, and the optic has, through constant correction of the olfactory errors, absorbed in its tracts an immense number of fibrils.

Correction of one sense by another is of momentary occurrence, most marked in awakening from sleep. The faint excitations of the registered impressions, through undue relativity of their action over the modes of impression while asleep, seem real to us, but after a startling dream we look about and *correct* the impressions aroused in memory which were taken for genuine ones, and which, by our surroundings, we now see were not so.

This constant antagonism of one sense by another would tend to enfeeble the erroneous impressions brought in, and the correction would entail the breaking down of nerve strands formerly operative, and the building up of new ones in the entire line of animals, on the basis of better adaptation ; and as these adaptations vary to such an extent that no two animals of the same species and no two persons are alike, then no two brains can be alike except in such matters as are common to all or to many. The musician's brain will differ in fiber distribution from the painter's, and the mechanic's from the laborer's. The civilized man will possess more " shunt " and discriminative diffusion fibrils than the savage. *And in precise ratio of intelligence of all animals do we see the cortical neuroglia encroached upon by fibrillar creation.*

There is a constant struggle on the part of higher strands to usurp the places of the lower ; and the reason for it is, that with the persistence of higher impressions in correction of the lower, the correction tendency builds up higher tracts toward lower planes. Were this to be fully accomplished, and no more impressions received by the animal than those to which perfect instinctive adjustment had been made, then consciousness would cease, and the animal become an automaton indeed ; but the eternal evolution and involution of things prevents this, and advance must occur or death, one or the other. The very conditions of our existence demand the perpetual change of adaptation to perpetually recurring change of environment. Such changes constitute the impression and cease to be experienced in consciousness as soon as perfectly adapted to. The rabble are for the most part thus adjusted, and tend to exhibit the automatic vegetative life with dullness of reason.

The environment which has reacted to arrange species identity has built up a common morphology ; a generic resemblance in brains exists, dependent upon a generic common experience and general resemblance of other parts made by the same law ; but when a family or succession of individuals live in a similar environment differing from that of the species, variations necessarily befall them, and inevitable retrograde or evolutionary changes proceed. This culminates in degradation on one hand and development of character or intellect upon the other, with concomitant change of brain minutiæ.

No sooner does the animal find itself susceptible to certain impressions that are of no life-conserving use to it, than the diffusion channels in the brain over which these impressions pass are exercised and consciousness evoked to the extent of producing *wonder*, an arrest, may be, of other motions, occasional doubts, shudders, tremblings, the eyes are opened wider to admit of better sight, a diffusion effect; and if at any time, either by accident or by persistent giving up to the exercise of the wonder impressions such impressions are turned to account, the after tendency will be to repeat the act upon revival of the impression, and step by step the useful, through association, becomes established and the useless held in check. But it does not always happen ; indeed, in but comparatively few instances does it happen that such integrations occur, and out of the countless billions of useless diffused impressions a few thousand cohere in this way, and out of this few thousand many, if not most, suffer dissociation through alteration of circumstances of environment or separation into new associations. For instance, an association may be formed on the supposition (mechanically wrought out line of least resistance) that a certain sound, scent or sight inures to the animal's advantage. This may be a gross mistake ; and suddenly it is discovered as such, through association with some fright experience. For thus the association begins to disintegrate, the fibrils set in new directions, and the last most powerful impression reassociates the strands, the thoughts, instincts, and reflexes.

Suppose a worm were to be often hurt by stones falling upon it, the peculiar tremors induced by the stone in falling would be associated with the memory of pain, and it would shrink thereafter upon the noise, even though no stone fell.

Association is induced, as often before remarked, herein, by simultaneity of impressions. If A and B arouse the feeling of pain, they cause a composite sensation which either A or B may arouse by this former simultaneity. We do not stop to analyze our sensations any more than the savage does. Two things occurring together *de novo* are likely to be accepted as part of one another, and association is the simultaneous sensation aroused, and is a simple term until subsequent dissociation analyzes the two impressions through a mental sense correction.

In the worm the associations were mostly tactile; next, feebly optic, with tactile and rudimentary auditory, olfactory and possibly gustatory, as far as Darwin's researches allow us to judge.

Analogy shows us that the social organism has the most primitive pain and pleasure ideas at first, acts upon alarms and jubilantly on the least occasion. It is also at war with itself, and so the units of the organism war with each other, except where their workings are the most obviously egoistic. The units are bound together blindly by their selfish instincts, as are the lowest natures. The cell works harmoniously with the neighboring cell from which it derives its food; the cat and dog say to the kitchen-maid who feeds them, as the child says to its mother, "I love you."

Altruism springs from and is a developed egoism. The organism which acts intelligently, altruistically, knows that it derives advantages from so doing which the lower, pure egoistic worker is incapable of understanding.

Consciousness diminishing in proportion to the outlet in activity for impressions is well shown when pain (dissonant cell motion) is assuaged by motion. In grief, one is impelled to walk about, move the arms, gesticulate, exclaim; and any motion affords some relief. When the pain is increased by moving, it is through the direct interference of the moving parts with nerve elements, and has no reference to such conditions as allow motion to allay pain.

CHAPTER VI.

PHYSICS OF THE CEREBRO-SPINAL NERVOUS SYSTEM.

The olden division of the senses into five must give way to a broader classification, and Tyndallian physics must be appealed to to widen our knowledge of the subject.

The reader should consult Herbert Spencer for an elegant description of the rythm and undulations which pervade all nature. He points out the up and down movements of atoms, molecules and grosser matter in light vibrations, sound waves, mechanical motion in the stellar universe, in plant and animal life, in sensation, in our feelings, emotions and reasoning powers. Everywhere we turn we find action and reaction. Individuals and nations at one time in the greatest turmoil, and at another time apathetic. One day we are capable of putting forth a great amount of energy, and upon succeeding days our faculties seem dulled, and so on everywhere.

These all-pervading motions constitute sensation whenever they affect the organism cell to the extent of causing motion in them. This may seem a sweeping claim, but it is justified in our inability to separate the psychic from the physical life.

In protozoa and metazoa without nerves it is plain that each cell presenting a modicum of its undifferentiated originality must act as a sensory cell to a greater or less extent whenever the molecules of such cell moved.

As the cell motions become coördinated by an evolved nervous system with higher and still higher integration and separation of cell labor into districts, the molar activities of the organism separated from the sensory, and while in every animal, however highly organized, the cellular senses remain as in the primitive form, such senses have nothing to do with those of the organism until internuncially related through the nerves. This may be better understood by regarding the condition known as anæsthesia. Where an injury to the sensory tracts occurs the parts from which such tracts proceed "lose their sensation." The animal, as a whole, is no longer apprized of what occurs in the division dissociated.

Modern physics afford us measurements of the motions occurring in such phenomena as light and sound, and glimpses, through analogy, of other energies.

Taking them seriatim we can best picture them as having wave lengths varying from the infinitely small to the infinitely large, even though we are unable to conceive of infinity. We can strike an approximate measurement for some forces and for others exact ones have been found for us.

Gravitation stands at the head of the list as acting upon every atom with a force inversely as the square of the distances and directly as the masses. La Place calculates the rapidity of gravitation to be at least seven million times that of light, for its interplanetary attraction is instantaneous, $7 \times 10^6 \times$ the speed of the violet ray per second, 757×10^{12}, gives us the enormous rate of 5299×10^{18}.

Chemical energy apparently merges into this from the fluorescent end of the spectrum, though we must make room for electricity as a vibratory force. Approximately where we cannot be precise, and exactly where science justifies us in our figures, we may tabulate the forces as follows:

Gravitation.............. $5 \times 10^{21} \pm$ to $+$
Electricity.............. $10^{15} \pm$ to $5 \times 10^{21} \pm$
Chemism..............$757 \times 10^{12} \pm$ to $10^{16} \pm$
Light..................392×10^{12} to 757×10^{12}

Heat	$10^8 \pm$ to	10^{12}
Sound.....................	40	to 4×10^1
Mechanical waves......	30	to —

But these forces interlace in many instances much more than can be shown in such a rude summary. For instance, fluorescence resides not only in the ultra violet ray with other actinic or chemical energy, but the latter extends into the spectrum toward the less refrangible rays. Heat extends into the ultra red ray. Sound and molar force have the same rates, grossly evident in low notes and less apparent in high ones. Then, to complicate matters still more, certain nitrites may be exploded by musical notes to which their molecules are disruptively attuned. We can only indicate these forces by such figures, and must leave a fuller consideration of them to the physicists. The main point to which I wish to attract attention is the somewhat serial vibratory nature of all the energies: gravitation, chemism, electricity, having wave lengths smaller and velocities greater than light. Heat and sound waves being longer and their speeds less than light rays, and the mechanical waves passing from invisibility in size to such monstrous dimensions as we can see them rolling on the sea, and beyond this to such as are produced by orbital perturbations of planets

Natural selection has adjusted the cell contents to certain vibrations; and as fast as evolution adapted these vibrations toward life-subserving processes, it mattered nothing in the end to be attained as to whether those waves were appreciated as sensations useful to the organism or accomplished some other and different method of perpetuating cell life, or whether, which is often the case, the same modes of motion were useful in two or several ways simultaneously. This is one of the many evidences of the impossibility of separating sensation from general molecular energy. We are compelled to select out such as are more plainly concerned in sensory operations, with the reservation that sensation is purely a relative matter in its divisions. It is hard to say where one sense ends and another begins. In the cases of touch and hearing in low notes and mechanical vibrations of about 30 per second, both senses appreciate the same thing.

Sir John Lubbock has demonstrated the ability of "lower" animals to be affected by vibrations ultra spectral and ultra acoustic. Man, thus, has either lost or never possessed faculties which other animals retain or may have differentiated. Individuals vary between themselves in their ranges. Most persons "have an ear" only for notes to 16,000 vibrations per second, while the possible range is placed at 38,000.

In default of observations which would have established the gradations as they arose, we cannot conveniently dwell upon the evolution of the senses except in a cursory way.

Light, heat and sound affect the protozoon, but with the gradual organ construction better defined methods for their appreciation arose.

Impressed with the fact that everything in our intellectual lives depends on or grows out of our sensations, David Hartley, a disciple of Locke, addressed himself to a study of the mental mechanism in the last century, when the anatomy and physiology of the nervous system were but imperfectly known.

The Aristotelian dictum which recognized the dependence of intellect upon the senses, until Hartley's day stimulated but few efforts to explain the intimate workings of animal machinery through direct study of its parts.

The surmises of Descartes and Hume were excellent, but Hartley's* inferences from observations of nervous physiology

---

*Observations on Man; His Frame, His Duty, and His Expectations, pp. 756. London, 1749.

are unparalleled, considering the crude knowledge of his time. His theory of vibrations was an application of Sir Isaac Newton's, whose personal acquaintance he enjoyed.

The sense organs and the nervous system are to be considered together. To Herbert Spencer we are indebted for a correct synthetical method of so doing, and however widely of the mark he may have fallen in many instances, through the necessity of his plan in seeking ultimates, the ground-work of Spencer's Synthetic Philosophy will inspire thinkers to the end of time.

In his data of psychology* he reviews the structure and functions of the nervous system, the conditions essential to nervous action, æstho-physiology, and the scope of psychology. We may advantageously epitomize some of the points he urges :

"While the rudimentary nervous system, consisting of a few threads and minute centers, is very much scattered, its increase of relative size and increase of complexity go hand in hand with increased concentration and increased multiplicity and variety of connections.

"Isomerism seems to be the molecular change undergone in nerve activity.

"Gray matter contains five times as many capillaries as white" (based on cubed averages from Kölliker's plates); from this he infers greater composition and decomposition in gray areas.

"Conditions essential to nervous action are :

"Continuity of nerve substance.

"Absence of much pressure.

"Heat kept above a certain level, limited.

"Suitable quantity and quality of blood supply.

"Minimum presence in blood of CO_2, urea, and other toxic agents.

* Principles of Psychology, Vol. I., Chap. i. to vii.

" Waste must be fully met by repair.

" Nerve is not capable of continuous stimulation or continuous discharge.

"The transmission of a disturbance through a nerve takes an appreciable time.

" Every wave of isomeric transformation passing along a nerve-fiber entails on it a momentary unfitness to convey another wave, and it recovers its fitness only when its lost molecular motion has been replaced, and its unstable state thus restored.

" Every part of the nervous system is every instant traversed by waves of molecular change—here strong and there feeble.   There is a universal reverberation of secondary waves induced by the primary waves now arising in this place and now in that, and each nervous act helps to excite the general vital processes while it achieves some particular vital process. The recognition of this fact discloses a much closer kinship between the functions of the nervous system and the organic functions at large than appears on the surface.   Though unlike the pulses of the blood in many respects, these pulses of molecular motion are like them in being perpetually generated and diffused throughout the body, and they are also like them in this, that the centrifugal waves are comparatively strong, while the centripetal waves are comparatively feeble.   To which analogies must be added the no less striking one that the performance of its office by every part of the body, down even to the smallest, just as much depends on the local gushes of nervous energy as it depends on the local gushes of blood.

" Higher animals possess greater self-mobility.

"Whenever much motion is evolved, a relatively large nervous system exists.

"Whenever the motion evolved, though not great in quantity, is heterogeneous in kind, a relatively larger nervous system exists.

"Whenever the evolved motion is both great in quantity and heterogeneous in kind, the *largest* nervous systems exist.

"Stimuli of all orders produce effects alike for like nerves."

We may safely posit what has here been quoted, the isomerism of nerve substance excepted, as needing more defence though satisfying many requirements. Few have shown the catholicity of Spencer in biological philosophy, and it would be well for every investigator to imbibe some of his spirit before advancing or condemning theories of mental action. Hume would rejoice in the immensity of psychological data we to-day possess, based upon dimensions and intervals of time, but which sadly need a Copernicus or Newton to assimilate them.

So far as objective phenomena are concerned, the vibratory theory affords the only practical and satisfactory explanation of universal phenomena to chemists, electricians, and physicists generally. Millions of dollars are daily invested by capitalists upon the accuracy of deductions from the vibratory theory in the mercantile fields of telegraphy, telephony, optics, chemistry, music, etc.

Schwalbe* affirms laws regulating the size of nerve-fibers. He found a physiological distinction, *e. g.*, motor or sensitive, influences the size of the fibers, and he adopts M. Pierret's law that the size of the cell and thickness of the nerve roots are regulated by the distance at which its nervous influence is exerted, corresponding to Dieter's claim that the size of the cell is in proportion to the thickness of the proceeding axis cylinder. The brain exhibits fewer size variations than the spinal cord. These variations, Schwalbe states, have reference to the size of the animal, extent of distribution, physiological attributes of the nerve, and the amount of usage to which it ministers.

Spitzka,† discussing the erroneous notion that function and cell-size were related, is willing to predicate for the so-called motor cell but one character, "that the transition from the body to the processes is so gradual that it is difficult to say

---

* On the Relations of the Calibre of Nerve Fibers. Leipzig, 1882.

† Journal of Nervous and Mental Disease. April, 1881, p. 325.

when the body ends and the process begins, while in unquestionably sensory cells the transition is always abrupt." Although in some areas sensory cells are larger than adjacent motor cells, as a general rule the motor cell is relatively larger than neighboring sensory cells; but, as Spitzka shows, this must be accepted with reservations.

Hoffman * affords us some careful measurements of nuclei, nerve cells and fibrillæ, but, in common with Spitzka, Ranvier, Meynert, and other exact observers, affirms nothing on the strength of mere size.

The nervous system does not differ materially from other tissues in development. Heterogeneity of function brings about a quantitative rather than a qualitative increase, and the only relationship between size and function, in my opinion, is a nutritional one. *Cæteris paribus*, thick axis cylinders afford less resistance than thin ones, and *use either in frequency or intensity, or both, would determine sizes between nerve-cells and fibrillæ of the same ages, lengths, locations, etc., without regard to motor or sensory differentiation.*

The great similarity between terminal fibers in their distribution to various organs, muscular, glandular, etc., admit of no other explanation than that the work accomplished by nerves does not depend so much upon the kind of stimulus or nature of the nerve elements as upon the character of, and possibilities inherent in, the non-nervous organs to which they are related.

The first appearance of contractile tissue left the primitive animal mass responsive to undifferentiated motions and sensations. Every jar or impact, every change of temperature or transition of light and shade, acted to stimulate the rudimentary muscular tissue. Adjustment to the environment determined to which of these stimuli response would be most apt to

---

* Journal of Neurology and Psychiatry. August, 1883.

occur, precisely such adjustment and readjustment as man in common with other animals undergoes daily. Definite sarcodal contractions were followed by the definition of a path for stimuli through least resistance lines.

Kleinenburg's neuro-muscular development of hydra Huxley[*] mentions as internuncial, and the primitive form of a nerve. Eimer[†] observed in the Ctenophora, the mesoderm traversed in all directions by very fine fibrils varying in diameter from $\frac{1}{30000}$ to $\frac{1}{11500}$ of an inch. These fibrils take a straight course, branch dichotomously, and end in still finer filaments, which also divide, but become no smaller. They terminate partly in ganglionic cells, partly in muscular fibers, and partly in the cells of the ectoderm and endoderm. Many of the nerve-fibrils take a longitudinal course beneath the center of each series of paddles, and these are accompanied by ganglionic cells which become particularly abundant toward the aboral end of each series. The eight bands meet in a central tract at the aboral pole of the body; but Eimer doubts the nervous nature of the cellular mass which lies beneath the lithocyst and supports the eye-spots. (*En passant*, this dichotomy of nerve-fibrils agrees with Schultze's observations.[‡]) Hubrecht[||] describes a new worm, "*Pseudonematon nervosum*, without trace of sexual, excretory, or sensory organs, with three muscular layers, consisting of a thick external longitudinal, a middle transverse or circular, and an internal longitudinal layer. The nervous system forms a continuous layer completely around the body, lying immediately inside the layers of muscular fibers. It consists of (1) a fine network of delicate filaments, appearing as if felted, barely tinged by the staining reagents, and (2) of scattered nuclei, belonging partly to connective tissue, partly to ganglion cells. The layer forms a continuous tube from the head, where there is no ganglionic enlargement, back through

---

[*] Anatomy of Invertebrated Animals, 1878, p. 62.
[†] Zoologische Studien auf Capri, Leipsic, 1873.
[‡] Archiv für Anatomie und Physiologie, 1878.
[||] American Naturalist, May 1884, p. 546.

the body to the caudal region, where the layer is present dorsally only."

Darwin's* ascription of intelligence to earth-worms does not involve the organization of a sensory distribution. Sounds produce waves in the undifferentiated sensorium, and alter the usual molecular interchanges. Light stimulates the cerebral ganglia of the worm without the intervention of a sense organ, owing to its diaphanous skin. Darwin finds evidences of olfactory and gustatory appreciation, but the tactile sense is the main one.

Gegenbaur† assigns a high development of tactile sense organs to vermes, but the visual organs of different species afford evolutionary variations. Those of the rotatoria are placed directly upon the cerebrum with a crystalline rod to each pigment spot, or the latter is present alone. Among cœlenterata, spongiæ have no sensory organs, neither have the low forms of acalaphæ. The cellular layer of the hydroid polyps is an undifferentiated organ of sensation.

" Balfour ‡ states that "embryological evidence shows that the ganglion cells of the central part of the nervous system are originally derived from the simple undifferentiated epithelial cells of the surface of the body, while the central nervous system itself has arisen from the concentration of such cells in special tracts.

" The nervous system of the higher metazoa appears to have been evolved in the course of a long series of generations from a differentiation of some of the superficial epithelial cells of the body, though it is possible that some parts of the system may have been formed by a differentiation of the alimentary epithelium.

"An early feature of the differentiation consisted in the growth of a series of delicate processes of the inner ends of certain epi-

---

* Formation of Vegetable Moulds through the Action of Worms, with Observations on their Habits, 1883.

† Comparative Anatomy : Bell and Lankester tr., 1878, p. 152.

‡ Comparative Embryology, 1881, p. 333.

thelial cells, which became at the same time specially differentiated as sense cells.

" These processes gave rise to a sub-epithelial nervous plexus in which ganglion cells, formed from sense cells which traveled inwards and lost their epithelial character, soon formed an important part.

" The central nervous system was at first continuous with the epidermis, but became separated from it, and traveled inwards."

Nerves, such as we find them in the higher types, originated from special differentiations of the nervous network radiating from the parts of the central nervous system.

Balfour admits two points as obscure :

1. " The steps by which the protoplasmic process from the primitive epidermic cells became united together so as to form a network of nerve-fibers, placing the various parts of the body in nervous communication.

2. " The process by which nerves become connected with muscles, so that a stimulus received by a nerve cell could be communicated to and cause a contraction in a muscle. The primitive relations between the nervous network and the muscular cells are matters of pure speculation."

It has often occurred to me that reasoning from analogies would afford clues otherwise unattainable. It is incontestable that the proper phosphorized protagonous composition was acquired by ingestion ; it lay in the environment of the low matazoon, and just as the shell of the snail, or more palpably, the envelope of foraminifera was excreted, so these explosive compounds were externally, ectodermally diffused, and, being acted upon by extrinsic forces in lines of least resistance, became useful in discrimination as the shell became protective, and we see every evidence of the tendency of animals to perpetuate and develop and differentiate organs which may have arisen, and in all probability did arise, accidentally.

The pigment spots in vermes becoming points of special irritation develop into eyes.

We can then assume that special chemical alterations occurring peripherally were appreciated centrally in terms of general sensation, that is, where previously jars and comparatively coarse waves served as stimuli, light vibrations, through chemical decomposition of pigment, brought about the primitive irritability constituting sight.

" The nervous system being derived from the epiblast implies that the functions of the central nervous system which were originally taken by the whole skin became gradually concentrated in a special part of the skin, which was step by step removed from the surface, and has finally become in the higher types a well-defined organ embedded in the subdermal tissues." (Op. cit.)

In the division of labor by the tissues, many portions fall below the average irritability by changes in chemical inaptitudes, and protoplasmic parts most exposed to vibratory impacts not only preserved the original sensitiveness, but developed it, as every organ is developed by use, perpetuating what is useful, and eliminating the useless.

Above the planula stage we have eminently contractile tissue aggregating in certain parts, and in others cells resisting differentiation in other than assimilative directions, the highly irritable ectodermal eventually forming the cerebro-spinal axis. At this period, internuncial fibers, such as nerves, arising in a plexiform manner, would fall into directions regulated by lines of least resistance, definite motions being the end toward which definite structures tended.   Muscular tissue is that structure which, preserving much irritability, developed its contractile powers, and became more efficient through linear arrangement of its cells and their connection with tissues of

some degree of induration. The contiguity of the highly sensitive tissue enabled possible conveyance toward the contractile of stimulation from distant parts of the body. This is the
neuro-muscular system *in futuro*. It is well not to mistake
words for things, and not designate nerves at this stage as sensory or motor, for the two functions seemed combined and inseparable. Vibrations are conveyed better by the rudimentary nervous tissue, and contraction of the rudimentary muscle
is the result. Such peripheral points as pigment spots,
through chemical susceptibility greater than had been heretofore apparent in the body mass, enabled the sensitive tissue to
take cognizance of new irritations, such as light waves, through
the conversion of such waves *into terms of general sensitiveness.*
An all-important consideration, and one whose elaboration will
assist clear ideas of neural differentiation. In the Morse code
of telegraphy, the dots and dashes arranged to indicate letters
are learned as indicating words or impressions, knowledge of
which was previously acquired. While the dot and dash
sounds were not new to the ear, their definite arrangement
brought us *en rapport* with a new experience correlated with
pre-existing appreciations. Sunlight existed and influenced
the protozoon, but as soon as the pigment appearance produced new terms for old sensations, a revival of former currents through a new influence, the recently acquired peripheral chemical change readjusted the life possibilities of the
animal, appreciation of light and shade became possible, *interruptions* of the molecular changes thus induced by association
with food acquiring or enemy escaping motions through use
and natural selection grew to the extended importance it possesses in metazoon existence. The pre-existing neural irrita
bility was not changed, but brought through peripherally
increased susceptibility into new relations. We find on every

hand a justification for this analogical reasoning. Whenever a new discovery or invention extends the range of man's abilities, it does not change the organism, but renders it capable of new applications of pre-existing potencies.

The rapidity of impact may modify subjacent structure so as to change the density of the conveying nerve, but the close resemblance of all nervous tissue precludes the idea of *any great change* being wrought in the nerve element proper through change in peripheral sensibility.

The translation into terms of nerve application must be peripheral, as it is initially there brought into relation with the former general sensibility. It is not necessary to consider the organ as subsequently dipping below, and becoming continuous from skin to central system, and mature differentiation shows us structurally that the centripetal nerves do not admit of this supposition. They may undergo slight modifications due to the times of impulses, but in the main do not differ from centrifugal nerves.

We have every reason to believe that the neuroglia is allied to the original undifferentiated irritable protoplasm, notwithstanding Huguenin,* "a tissue which increases as mental functions decrease cannot be the medium of such functions," for as differentiations of this basis substance occurred in a higher scale of intelligence, it is the very substance of all others to be encroached upon by organization.

The contractions and expansions of masses through changes in density by aggregation and seggregation of molecules are familiar chemical reactions. That identical molecular translocations constituted the irritability and automatism of living protoplasm we have every reason to believe.

The development of the aggregating and seggregating in-

---

* Allg. Pathologie der Krankheiten des Nervensystems, Zurich, 1873.

stability in either direction would inevitably cause a separation
of such over susceptible molecules from the general mass.
The eminently aggregative molecules are the muscular. The
eminently seggregative molecules are the nervous, and both of
these are developed from the irritable protoplasm, which as
the neuroglia separates off into ganglionic areas, and through
preservation of the original irritability (antagonistic seggregative and aggregative molecular power) and elimination of interfering matters, it has become the seat of the feelings, the
meeting place of the sensations, the part to which waves converge, and from which they diverge in the institution of vital
movement.

The general sensibility which guides the muscular contractions when the central nervous system became the conveyor
of general sensibility, its connection with the muscular system
would follow, and whenever exaltation of sensibility occurred
in any area, centripetal connection with the seat of general sensibility would also follow. We can safely grant the primitive
central neuroglia, such as the amphioxine, a general susceptibility with a fixed range of mechanical and chemical irritability, the gelatinous quivers and molecular motions which *necessarily* occur as conditions of its existence.

The motor nerves respond to a definite group of these vibrations through habit and adjustment. No matter how these
vibrations may be produced, we know that when they occur
they stimulate the muscles to contraction. The uniform range
of vibrations within the compass of the protoplasm producing
muscular movements necessitated the removal of motor nerves
from chances of passing stimuli, and hence the motor nerves
in highly organized animals are related to the central nervous
system, but not to the periphery. The advantages of this are
apparent. The limited field of disturbance to which the cere-

bro-spinal axis or the protoplasmic basis substance was subject served the motor nerves as a cut-off for impressions.

The somatic motions are innumerable, but may be grouped
as molar and chemical. The molar motions are the circulatory, and those made by the limbs and body generally in locomotion, prehension, or any other change of relative position of
parts of the body, regarded by Goethe as comprehending all
that animals could do.

The atomic interchanges which occur in every cell of the
body produce chemical movements of complex natures. Respiration, in the conversion of hæmoglobin into oxyhæmo-
globin, is chemical, and in muscle and nerve this respiratory
function is evident. Oxygenation of the tissues is an accompaniment of life in all cells. It is a nutritive act, and the significance of oxygenation is no greater in the nerve areas than
it is in other organs. Hence oxygen, though necessary to
nerve action, does not impart that action, but contributes to it,
renders it possible.

The multitude of molar and chemical motions in the body
produce a great number of vibratory waves in all tissues.
Differences of temperature between parts induce circulations of
heat waves. Differences in acidity and alkalinity with osmosis
doubtless induce electrical changes, but none of these movements can be said to be the nervous force. If we consider the
nerve elements as unstable and readily stimulated into activity,
we cannot ignore the myriad stimuli of every movement of
life, awake or asleep.

It may be granted that electricity is one of the energies under consideration, within definite limits, as is the case with
varying degrees of heat, with which the body in general and
nerve elements have established adjusted relations ; and before

proceeding further, it would be well to examine this perpetually recurring matter of adjustment and readjustment.

We experience impressions, sensations, resolvable into vibrations, which are novel and provoke reflex or other motions. Often repeated, these vibrations cease to produce the first effect, "we grow accustomed to them" suffices generally as an explanation, but to the physiologist that should be no explanation. The only inference to be made from analogy is that in this failure of repeated stimuli to produce a repetition of the effect, a readjustment on a physical basis has been made. A change in the arrangements of the molecules which were at first influenced has been effected. The irritation can be made to induce its first effect, but a *diffusion* of the vibration into other and more general channels has evidently occurred. A *change of plane* is made, and nothing but a rearrangement of the tissue over which these impulses pass can explain this adaptability of the organism to changes in environment.

Under repeated strains, muscles visibly adapt themselves to new demands by growth. The nerves are no exception to the rule of tissue-increase through exercise, but it would be a narrow view to assign growth of nerves as the only means of adaptability, for demonstrably induration of other parts and diversion of impulses into diffused channels operate toward the same end.

The physical basis of readjustment which enables the miller to sleep undisturbed by the noise of his machinery are of this complex nature; readjustment of many tissues stimulated to simultaneous growth or other change, as the tympanic membrane changes, the tensor tympani relaxation and alteration in blood-vessels. This is more apparent in the artillerist and boiler-maker, whose readjustment entails obtuseness to other and ordinary vibratory influences. The rustic is excited by the confused noises which the city-dweller does not notice, and the latter feels oppressed by too long a continuance of the quietude to which the former is adapted. It is safe to infer that no two nervous organizations are precisely alike, and that ranges of difference exist in the same species, even rendering brains unlike in minute distributions and texture, when they may present the same gross morphological appearances.

The general molecular changes which normally go on in the nerves are readjusted to an irritative influence, and discordances are inevitable; a slight noise or jar suffices to induce reflexes which were before unexperienced, and this indicates, in common with many other phenomena, that interruptions to the ordinary vibrations afford the basis of reflexes and sensations.

An excellent illustration of my interruption idea is one afforded by respiration, which we do not ordinarily feel or of which we are unconscious. The neural changes are constant, regular, rhythmical, but when a change of plane, such as relative increase or decrease of sensation or motion in other parts, or diffusion, or a shock of any kind interrupts the motions of the heart or lungs, forthwith sensation is evoked, and the change is realized or felt.

Both the telephonic and telegraphic workings afford analogies. The wires are traversed by molecular vibrations which exert no influence over the relay armatures and sensitive diaphragms at the stations, but *interruptions* to the passage of these molecular movements instantly become apparent in the terminal apparatus. Such interruptions may be effected by solution of continuity, increase or decrease of current temperature, or hygrometric variations, etc.

The nervous system, as well as the organism itself, is being perpetually readjusted to its environment, and no sooner is one plane reached than alterations occur which may act as stimuli, but cease to so act. It is only through alterations of adjustment that impressions are possible. These plane alterations are, of course, limited by chemical possibilities related to the integrity of the tissues.

The following matters bear upon this change of plane: .

Muscles which by degeneration have lost their nervous supply are more responsive to the constant current than to induction shocks. Thus the condition of nerves in cases of paralysis may be tested.

Nerves suddenly heated or cooled induce muscular contractions, but not when application is gradual.

Moderate warmth, 45° C., in frog, favors nerve and muscle activity.

Cold at 0° C. diminishes the nerve force from 28 m. to 1 m. per sec.

" Personal equation " shows the varying sensibility of individuals and between them.

Helmholz fixes the rhythmical discharges over motor nerves to contract muscle at 18 to 20 in most animals, and 16 to 18 per sec. in frogs. Löven found these discharges to be 12 to 13 per second in a man's arm.

The law of contractions, analectrotonus and katalectrotonus are interpretable as increase from low to normal, or from normal to high planes. Ritter's "making and breaking tetanus," shows how convulsions originate in heightened or lowered vital activities.

All show variability of nerve vibrations within a range. "Regarding sensation as the sum of a series of increments of sensation corresponding to increments of stimulus, Fechner adopted the mathematical operation of integration, and concluded that sensations increase, not in proportion to the strength of the stimulus, but to the logarithm of the strength of the stimulus."*

---

*Ross: The Diseases of the Nervous System, Vol. I., p. 83.

A change of plane changes effect of stimulus. Hyperæsthesia is an exalted plane, anæsthesia a depressed plane.

Adding algebraically stimulation to the planes affords good numerical results; thus, let the normal plane equal 1, the hyperæsthetic 2, then a normal stimulus of 1 added to normal plane equals the ordinary reflex 2=hyperæsthetic plane; but $1+2=3=$pain, which is an accompaniment of all transgressions of usual limits.

Some partially deaf persons hear best amidst great noise.

The law of relativity includes the change of plane consideration. This law, of such universal application, is well discussed by Bain*.

He shows that it must be perpetually kept in view, in discussing the emotions, that the measure of a feeling is the measure of a transition.

Pain and pleasure are purely relative.

Change is necessary to feeling.

We are unconscious of unremitted impressions.

The degree of feeling is proportioned to the change.

Abruptness or suddenness of transition is one mode of enhancing the effect.

Most air-breathing animals are insensible of the pleasure of a transition from confined to pure air, as they are not so apt to experience it as man is.

Fishes, in tropical seas, are accustomed to uniformity of warmth. Never to feel cold is never to feel heat. Hence we are unconscious of what is not felt.

Sightless animals have no sense of darkness as we understand it.

Pleasures and pains often depend upon the one preceding the other.

Many pleasures only become such by their cessation.

Any excessive indulgence loses its zest.

Novelty is often pleasureable, and first experiences are most acute, as in puberty, and when traveling, learning, etc.

_____

*Bain: The Emotions and the Will, p. 57 et seq.

The longer the remission, the greater the stir or shock of renewal.

After a long privation, a pleasure may regain very nearly its pristine charm.

The perfection of enjoyment in anything is a mean between the repetition that makes the pleasure stale and the privation that leaves an aching void.

The serenity of the man of business under shocks and vicissitudes, opposition and abuse, is, in part, though not entirely, due to their frequency.

Case-hardening through familiarity with suffering is inevitable.

Ennui is the feeling during removal from usual variety of pleasures.

Monotony and painful sensibility succeed exhaustion of organs after stimulation.

Happiness depends upon rotation of familiar pleasures.

Established nerve impulses cease to affect, but interruptions to them are appreciated.

The law of diffusion Darwin amplifies in his Expression of the Emotions in Man and Animals, and Bain regards it as carrying with it the law of relativity.

According as an impression is accompanied with feeling, the aroused currents *diffuse* themselves freely over the brain, leading to the general agitation of the moving organs, as well as affecting the viscera.

The organs prominently affected by diffusion of nerve influence are the members, especially facial features, ears in animals, next the viscera.

Pain may be suppressed by excitement.

The senses are correlated so that one feeling *must* predominate at a time. The eyes, ears, and hands cannot minister to three different feelings at the same time. Frequently a shock of feeling will stop a walk.

No emotion but that it has outward display, if intense enough.

When feeling is weak, expression is weak.

Education restricts impulses to definite channels. The man no longer sticks out his tongue, as the child does, when he writes.

To recapitulate:

Eminently contractile tissues (muscles) and eminently irritable or mobile tissue (ganglionary gray) underwent differential development and separation. Eminently sensitive organs were subsequently developed from indifferent substances (mainly pigment) in the ectoderm.

These three groups were united internuncially.

All developed from the primitive fibrils arranging themselves in lines of least resistance; such fibrils as conduced most to useful action being perpetuated and developed through use, the remainder of the network remaining undeveloped. In this way, definite nerve tracts arose, grew, and were inherited, undergoing subsequent development and alteration through the operation of the same laws which originally called them into existence.

Nerves undergo mainly quantitative development, *i. e.*, sizes and densities of axis cylinders may differ, but nerve tracts are essentially alike.

End organs differ qualitatively.

Increase of connections is followed by increase of function differentiation; and where commissural union is greatest in brains, intelligence is greatest (Spitzka *).

Sizes of axis cylinders and nerve cells are regulated by use.

Time is required in all nerve phenomena.

Nutrition, repair of waste, the assimilation of protagonous material from the blood, as well as oxygen therefrom, is a prime essential to nerve-action. Exhaustion for shorter or

---

* Architecture and Mechanism of the Human Brain.

longer time follows action, denoting something consumed, and gray matter is the place where the greatest consumption of material occurs.

The necessity for the absence of much pressure points to a mechanical factor in nerve-mobility, as well as a chemical; a certain freedom for molar motion is a requisite.

The increased amount of force put forth by an animal increases nerve size.

The multiplicity of motions increasing, increases size of nervous system.

Both quantity and multiplicity of motion increase it to the greatest extent.

Nerve fibrils are in constant motion, molar or molecular, or both, during life, independently of the atomic and molecular motions involved in the disintegration and reintegration of all physiological structures.

Indifferent tissues may and do convey vibrations, and, when passing centripetally over other than nerves, produce central nerve action, though the cushioning of the cerebro-spinal fluid minimizes this. Nevertheless, extraneous vibrations, such as shocks and jars, are apprehended centrally; any way in which such jars could affect the centers, would become the mode in which such jars are cognized. A fall upon the head inducing flashes of light and tinnitus aurium show, in common with other matters, that molar motions of the coarsest kind can induce the molecular movements of sight and auditory centers. The shiver produced by unpleasant vibrations experienced in riding over badly-equipped railroads, or in some by passing the hand over velvet, are other instances of the interruptions to normal waves sufficient to provoke diffused reflexes being generated by ordinary molar motions. Even the horripilation felt in handling a peach is of a mechanical nature.

9

All nerves convert their initial impacts into terms of general sensation, as evidenced by homogeneity of nerve-structure.

Interruptions to and changes of normal nerve-currents constitute stimuli.

Adjustment and readjustment of the nervous system is incessant and causes change of plane of nervous action in varying degrees.

Change of plane changes effects of stimuli.

Fechner's law of stimuli increase is involved in the foregoing considerations as a sub-consideration.

The laws of relativity and diffusion are all important in connection with nervous and mental operations.

These matters being conceded, the impossibility of assigning any special rate of vibration or wave length for normal nerve impulse can be seen. Only the broadest generalizations are admissible, such as warrant the belief that nerve action is more rapid in birds than in mammals, swifter in the latter than in reptiles, and the interruptions to nerve action would produce quicker reflexes in such animals as possessed the quickest nerve currents. The existence of constant currents is a necessary outcome of the acknowledgment that nerve molecules move at all, and that there are ever present sources of stimulation.

Helmholz assigns twenty-eight m. per second as the rate in frogs, thirty-three m. in man. Foster * adopts the latter rate for both sensory and motor activity, believing himself justified in so doing through inability to otherwise reconcile the various discrepancies of investigators. The normal current would vary in health and disease; probably the most accurate figure for the rate could be obtained by using the circulatory as a multiplier. The pulse-wave velocity of between nine and ten

---

* Text-Book of Physiology, Reichert's-Foster.

m. per second (Weber) could be taken as a factor, though the widely-varying velocity of the blood itself in different channels and its differences of pressure are potent as increments of neural velocities and are results often of such velocities.

Adjustment has taken place with relation to the circulation so that it is seldom felt save in overplus rapidity and in the states which produce tinnitus aurium.

The relative swiftness of pulse and nerve waves stand thus as one to three or thereabouts, often rising and falling together or reacting upon each other.

Helmholz' 33 m. per second suffices to base calculations upon in all nerves.

Schmidt * measures fibrillæ of neuroglia at $\frac{1}{100}$ to $\frac{1}{800}$ mm. in diameter, and axis cylinders of adjacent cell processes at from $\frac{1}{200}$ to $\frac{1}{800}$ mm.

Thirty-three m. per sec. with axis cylinder composed of globules .01 to .001 mm. in diameter affords an impact or explosion between globules at the rate of from 33×10^5 to 33×10^6. Let us adopt 10^7 as the usual rate, for convenience of calculation, that is, as the rate common to some nerves under some circumstances.

But the actual rates of nerve molecular motion are of as little consequence as those of any other chemical change. What value could attach to a knowledge of the rapidity of disintegration and reintegration of the water molecules going on in a jar of battery in electrical generation upon Grötthus' hypothesis?

Haycraft † in an essay upon the limitations of time of conscious sensation, states that:

If any sensory surface be stimulated for a given period, a

---

* Journal of Nervous and Mental Disease, January, 1879.

† Brain, April, 1884, p. 141.

sensation will be produced which will not exactly correspond with this stimulation in point of time.

The sensation will be produced a certain interval after the first movement of application of the stimulus. It follows from this that two stimuli following one another sufficiently rapidly will give rise to a single sensation. The correspondence in time between the stimulus and sensation varies in the case of the different senses.

Tactile impressions applied slowly (one to forty per second) give rise to distinct and separate feeling. More rapid stimuli are fused and a feeling or sensation of "roughness" is produced.

If the impacts follow faster than fourteen hundred per second, the sense of "roughness" disappears and a sensation of single impact is produced. Tickling is produced whether an impact stimulates a number of sensory areas one after the other, as in drawing a feather over the skin, or if impacts follows one another on the same sensory area; for impacts of fifty to fourteen hundred per second, if applied to the same sensory area, produce "roughness" if forcible enough, but when lightly applied they produce "tickling."

Different areas of the skin vary in sensibility. Sensations of heat and cold vary in times of appreciation. Haycraft holds that the differences in the exact limits of the feelings seen in the different sensations are not due to any difference in the centrally produced feelings, but to the differences seen in the translation of the external energy into nerve energy through the end organs.

In the sensations of sound and light, as well as in the tactile sensibility, when stimuli are repeated more and more rapidly, a period preceding complete fusion is reached when the sensation produced is disagreeable in its nature. In the case of sound, the irritation is produced by beats; the cause of "dissonance" in music is due to recurring stimuli not to be separated in consciousness, and yet not rapid enough to give rise to a simple sensation of sound.

The flickering of a flame is irritating in its nature, and tick-

ling in the case of tactile sensibility may give rise to excrucia-
ting agony. We may compare, therefore, roughness or tickling, tactile sensations, to dissonance in music. Bloch* finds
tactile sensibility from finger of hand took $\frac{1}{4}$ second longer
than by auditory apparatus. Vision is most rapid, next sound,
which is $\frac{1}{2}$ second longer than visual transmission, and tactile sensation is $\frac{1}{4}$ second longer in transmissions than vision.

If nerves conveyed physical forces *per se*, there would be no
use of end organs. It is in these peripherally located organs
that translation of impacts occurs into terms of interruptions of
previously existent vibrations.

Stimuli added to the ever-present stimuli which induces the
normal can but change the rate of speed, but when once
changed and persisting, it becomes a change of plane and
ceases to be a sensation. It is, therefore, the rise and fall of
rapidity which is appreciated, and the rapidity of this rise and
fall is the thing appreciated, for slowly induced changes we
know are unfelt.

We cannot multiply the number of vibrations causing the
sensation into the pre-existent waves, for the reason that the
new effects of stimulation, if prolonged, become blended with
the previous normal. An acceleration of the normal must be
granted as an effect of all stimulation, but continued stimulation ceases to accelerate, and a plane near the former normal
is reached in longer or shorter time. Thus the "palling" or
cloying of senses occurs.

Sensation, so far as movements among the nerve elements
is concerned, is accompanied by two cond'tions, the temporary
acceleration of the normal, and intervals between impacts of
acceleration. The normal can, by exhaustion, drop below its
usual plane, so far as to, though relatively, cause the sense of
an interval between impacts.

*Gazette des hôpitaux, No. 128, 1883.

Conceive 10^7 the nerve normal.   A single impact in a second raises this.   The instant of contact, the acceleration and the cessation or drop to normal, which may not be simultaneous with duration of impact, constitute the impression.   The problem would be simpler if simultaneity existed, but as the duration of an impression outlives the impact, it must be duly considered.

Tactile fusion occurs with fifty and with fourteen hundred impacts per second.

Ocular sensations, repeated ten times a second, fuse.

One hundred and thirty-two dissonant "beats" of tuning forks fuse.

Forty auditory vibrations per second last $\frac{1}{11}$ second, forty thousand last $\frac{1}{200}$ second.

Residual optic sensation for light of moderate intensity is $\frac{1}{5}$ second.   Red, violet, and green last in the order named.

The interval between optic stimuli varies according to the light intensity, being shorter with stronger light.   Faint light has an interval of about $\frac{1}{10}$ second, strong light $\frac{1}{30}$ second to $\frac{1}{20}$ second.   Though the sensation is longer with the stronger light, the decline begins earlier, and successive sensations of bright light are fused with greater difficulty.   The interval at which fusion occurs is shortest with yellow, intermediate with red, and longest with blue.

The duration of a stimulus necessary to affect the retina is exceedingly short, as when the electric spark causes sensation of light.

The smallest difference of light which we can appreciate is about $\frac{1}{100}$ of the total luminosity used.   The same law holds good with the other senses.   The smallest difference we can detect in length between two lines is the same fraction of feet or inches, in lines feet or inches in length.   Weber's law is, that appreciation of stimulus increase varies in proportion to the whole stimulus, and it fails where the stimulus is very small or very great.   Fechner's law is that the "sensation varies as the logarithm of the stimulus."

According to the obtuseness or extreme sensitiveness of the end organ, so will the duration of the impression be shorter or longer.

For the same organ slow vibrations are felt longer than quick vibrations.

For the same organ intensity of impression lengthens the duration. But the rapidity of conduction is a different matter. This has been worked out through Exner's "reaction periods," but we cannot use the tables, for the time of conversion of impressions into nerve waves and motor effects are not separated. To determine differences between senses it might answer, but not between muscle and nerve action.

The fact, however, that this action is accelerated in hyperæsthesia, and retarded in anæsthesia, bears upon the change of plane theory.

There is obviously less resistance to the tactile movements in nerves admitting full appreciation of forty impacts per second as separate sensations, and imperfect appreciation of these to fourteen hundred per second as "roughness," than in the optic nerve movements, which fuse impressions of ten per second.

The question may well be asked, Is this due to continuous peripheral action or to the central duration of the impression? "Dilemma" times and durations measured on nerves separated from terminals, suffice to show it to be largely non-peripheral, though there is a mixture not eliminated.

In an endeavor to trace the evolutionary aspects of the different senses, it is to be remembered that the tactile sense itself has undergone development, and that the primitive sensibility is far more obtuse. The greater duration of an impression is, in some respects, a measure of its higher development.

Though both tactile and auditory are closely related, the manner in which the receptive organs are impressed in both cases differ. The cuticle is directly touched with a comparatively hard substance in one instance, and in · the other an elastic medium, the air, conveys its pulsations to the tympanum, and modifications of these waves are undergone before the auditory nerve is affected.

Mayer's researches on the male mosquito antennæ prove in that case that the aërial pulses affect the fibrils at right angles to the antennæ ; the latter conveys the resulting waves inward, but there is a multiplicity of arrangements for effecting auditory appreciation between different invertebrates. The vertebrate auditory apparatus has a decidedly mechanical structure.

The ability to distinguish separate impacts in time is one thing, and the ability to distinguish series of vibrations apart from each other is another. The first is of a molar nature or moleculo-molar, the second is molecular and consists of chemcal disruptions comparable to the explosion of the nitrite when the musical note to which it is susceptible tears its molecules apart.

Helmholz * says : "As the difficulty of making a trill in the bass is the same in all musical instruments, the vibrations of the mobile parts of the ear for bass are not damped enough to prevent two sounds succeeding each other so rapidly without blending. There, hence, must be in the ear different parts which are set in vibrations by sounds of different heights, and which give the sensations of these sounds.

"For 40 vibrations per second, a residual sensation lasting $\frac{1}{4}$ sec. For 40,000 vibrations per second, a residual sensation lasting $\frac{1}{500}$ sec. Thus the residual sensation of 30 vibrations per sec. should last $\frac{1}{10}$ sec., for they follow at $\frac{1}{30}$ sec. intervals. Why do they blend?

---

* Tonempfindungen, p. 215.

" Do not these distinct impulses fall on the ear in $\frac{1}{6}$ sec.? Co-vibrating bodies in the ear tuned to vibrations below 40 per second do not exist. The ear vibrates *en masse*, and the duration of these oscillations of the ear as a whole are far too short to remain the $\frac{1}{6}$ sec."

Some of the agencies to produce vibratile motion are comparatively coarse, such as otoliths, while the tactile corpuscles receive their impacts direct, and their wave interruptions appear to be of a simpler nature; the auditory apparatus is elaborated to converge the diffused sound waves, and convert them into vibratile motions of lesser terms. The lower series of notes act as tactile impressions to the ear and induce moleculo-molar changes, the higher up to 40,000 (the range being only to 16,000 in most persons) waves per second produce such neural motions as are closely allied to the 1,400 per second " roughness " discrimination of the tactile sense wherein the molar motion is a minor product and the molecular the major one.

The moleculo-molar rapidly fades into the molecular as 40,000 is approached in hearing and 1,400 in feeling. The disagreeable feeling Haycraft notes as preceding fusion is the dissonance of erratic nerve action. The capacity of the nerve for certain rates of waves is transcended, and correlatively painful mental feelings occur during doubting or constraint. Higher than 40,000 auditory waves are not appreciated, because there is no end organ in man rendering him susceptible to them.

The opacity of nerve substance precludes the possibility of light vibrations passing to the cortex. Heat appreciation must be translated into lower terms, for high heat molecular movements are disintegrative. We are forced to the conclusion that neural vibrations are normally higher than sound, over 40,000, and much lower than heat vibrations of the ultra red ray.

The disintegration of the visual purple, causing interruptions of optic wave-lengths, seems the most rational explanation of sight. Rhodopsin absorbs all rays, and it is fair to presume that upon its disintegrative times will depend the color of the light perceived.

Gamgee* dwells upon the bleaching phenomena, and cites many facts which could be usefully considered in elaboration of a chemical interference theory. We have room for a few only.

The order of bleaching is yellowish-green, green, blue, greenish-yellow, yellow, violet, orange, red.

According to Boll, light perpetually destroys the retinal color, and darkness regenerates it. Thus in the space of a wink regeneration occurs, and light, acting upon many points, affords time for regeneration of points acted upon. Mayer† appropriately remarks: "May not research in this direction be guided by the hypothesis that the molecular constitution of the retinal rods and cones is such that their molecules are severally tuned to vibrations corresponding to the red, green, and violet? This would lead us to look for effects of actinism on the retina, as showing the link existing between the transmitting and sensory functions of the eye. Do not the facts of the known persistence of chemical action, after it has once been initiated, and the time which would be required for the retinal molecules to re-combine or re-arrange themselves after the ethereal vibrations had ceased, comport with the known durations of the residual visual sensations, and with the main facts of physiological optics, better than the hypothesis that masses of the retinal elements are set in vibration, rather than their molecules?"

Actinism is often erroneously used in the sense of being the only chemical force. Many forces produce chemical effects, such as sound, heat, electricity, as well as light, but actinism is

---

* Physiological Chemistry of the Animal Body, p. 465.
† Researches in Acoustics, Am. Jour. Arts and Sci., 1874, p. 251.

the light chemical force, and abounds in the violet ray and beyond, though inherent in all rays. The molecular force mentioned herein as constituting the actinic is of lesser wave
length than any light ray. White light appears to me to be
the resultant of all rays through destructive impacts, while
purple absorbs all rays. The law of relativity causes the eye
to accept as white light any light prevailing if intense enough.
So the nervous system growing habituated to myraid impulses,
accepts these as the normal ones, and variations from the plane
adopted become the excitants. Many noises to which the ear
grows accustomed constitute an analogous datum plane for
audition, which must be transcended in amplitude or wave-
length to stimulate the acoustic apparatus.

Color perception is a differentation of simple perception of
light and shade, such as we have reason to think constituted
the rudimentary light perception. The subsequently acquired
media of the eye act as does water in cutting off the heat rays,
otherwise heat would be seen. A pigment which responded
indifferently well to light, gave place to the visual purple with
definite actinic attributes. From the great rapidity of all light,
there must be an intensification of pre-existing nerve vibrations. In the dark, the normal vibrations occur over the optic
nerve, but a little light suffices to start the molecules to more
rapid interchanges of position. Periods of vibrations, or times,
constitute pitch in music and hue in color. The amplitudes
decide the intensities of sound and the brightness of light.
Hence, in dealing with stimuli, the height as well as length of
a wave must be measured.

The rapidity with which rhodopsin changes, varying in different individuals, culminates in extreme variability in color-
blindness, and the acuteness of color perception which accom-
panies or precedes some forms of insanity. Sight is dependent

upon visual purple interference or absorption of rays. Where the pigment fails to bleach, or changes too rapidly or too slowly, color aberration would be inevitable.

M. Morren,[*] referring to Tyndall's Kinetic Energy of Vapors in Light, says: " If a body forms and maintains itself in certain undulatory conditions, it is necessary that the oscillations of the atoms which constitute its molecule should be different from those of the medium where the body is produced. But if the body is transplanted into another medium, where vibrations synchronous with those of its atoms are produced, the vibrations of these last become more energetic, and the live force, which they accumulate, thus becoming considerable, the atoms are thrown to a distance from each other greater than the radius of their sphere of action.   The atomic edifice, previously formed, is demolished, the atoms preserving their special attractions for a new edifice, possible in the conditions of oscillation which surround them, consequently not possessing longer the same synchronous oscillation as those of the medium."

Thus, with the free light waves saturating rhodopsin with wave lengths .0007 mm. long, at the rate of 392×10^{12}, for red light, and .0004 mm. long, with a rapidity of 757×10^{12}, for violet light, retardation of these times and elongation of the waves takes place, raising our hypothetical nerve-current from 10^7 to x, y, z $\times 10^7$. The multipliers can readily take the place of the three nerves of the Young-Helmholz theory of red, green, and violet light, and intermediate rates would account for the other colors, the x, y, z being molecular modes possible to the optic tract and synchronous movements of any two or all constituting the composite color sensation.

The increased rapidity of nerve action, with increments of heat, about 1 m. for 3° C., is an expression of the physicist's

---

[*] Comptes Rendus, Aug. 9th, 1869.

observation that the "kick" of a molecule depends upon its increase in temperature.

Lockyer compares the motions of sound and light to the orderly movements of a company of soldiers, and likens the molecular motions induced by heat to the irregular movements of a crowd. This simile may be carried into neural vibrations, those induced by electricity, light, sound, or tactile impression being orderly and definite, while heat tends to add to the instability of the molecule, and where it induces movements, they are of an irregular nature; but these very irregularities constitute their mode of impressing themselves upon the consciousness.

Olfaction apparently depends upon an irregular molecular motion of particles impinging upon the Schneiderian membrane distribution, odorous substances having regular orbital rotations constant for each substance, aerial interferences making their conveyance to the nostril irregular.

Gustatory sense, when separated from olfactory and tactile accompaniments, evidently stimulates the taste buds through solution and chemical disruption, rotations, and impacts. Above and below 40° C., taste is impaired, showing the circumscribed nature of this differentiated tactile sense.

Touch......... $= 10^7 \times$ b $+$, Steady pressure to 50 and 1,400 impacts per second.

Taste........... $= 10^7 \times$ c $+$, Moleculo-molar solution impacts.

Smell ... . $= 10^7 \times$ d $+$, Molecular rotations

Hearing....... $= 10^7 \times$ e $+$, 40 to 40,000 aerial pulsations per second.

Sight........... $= 10^7 \times$ f $+$, Rhodopsin bleaching by atomic vibrations 392×10^{12} to 757×10^{12} per second.

Heat............ $= 10^7 \times$ g $+$, Irregular vibrations with amplitudes and lengths increasing 1
m. per 3° C. per second.

The law of diffusion may assist in seeking the values of a to g.

Some effects of diffusion upon nerves are secondary, such as those resulting from the heart palpitations, tremblings, flushings, etc.; but primarily diffusion may increase or decrease general sensibility, tactile in production of formications, gustatory sense in producing taste impressions (occasionally during fright, anger, or electrical stimulations of other nerves, showing close relationship of taste and tactile), vaso-motor and motor disturbances. Sometimes tinnitus is induced, but this may be secondary from vascular changes. Audition, olfaction, and sight are in health generally beyond the range of diffusion of nerve excitement.

For working purposes, arbitrary values of a to f may be taken as ranging from 2 to 7. Heat having an altogether different motion from the other forces, g would have hyper-neurotic general value to be considered apart.

Thus the higher molecular energies, as electricity, light, heat, and the moleculo-molar energy, sound, may stimulate sensations of each other in a descending series, and produce tactile and motor phenomena, but, as a rule, the lower rate of vibrations cannot produce the higher, except in abnormal conditions of the nerves when hallucinations of the higher senses are possible.

This explains why paræsthesiæ are so common as to pass almost unnoticed, but aberrations of the higher senses denote profounder disease.

The loss of taste and smell are the most common affections of special sense associated with anæsthesia; loss of hearing is less frequent, and loss of sight is rare. In hemianæsthesia, taste and smell are usually abolished on one side, but hearing and vision are only diminished.*

When a locomotory organ arose in metazoon existence, there was a concomitant necessity for a definite amount of nerve vibrations to pass to that organ; otherwise it would either not act or would react to all stimuli. Thus a subsequently acquired organ must have another amount of impulse over its nerves. These rates must be relative to the plane of whole neural activity. Effort means increase of molecular activity, and as weighty organs require more muscular effort, so there cannot fail to be established differences between *amounts* of neural vibrations sent to parts requiring to be moved.

In telegraphic engineering, the *amount* of current evolved is termed its quantity, which may be increased or diminished by lessening or increasing the resistance in circuit both in the generator and on the line. The quantity of electricity which in any unit of time flows through a circuit is called the *intensity* of the current.

This intensity is equal in all parts of the circuit, no matter how heterogeneous the parts are, and it is proportional to the electro-motive force.

Resistance is inversely proportional to the intensity of the current. From this, Ohm deduced his celebrated law: "The intensity of the current is equal to the electro-motive force divided by the resistance," which is expressed in the formula: $I = \frac{E}{R}$ where I = intensity, E = electromotive force, and R = the resistance.

Many telegraphic laws are applicable to the nervous system. We derive four considerations: intensity, quantity, resistance, and energy (molecular). The differences between quantity and intensity may be apprehended by comparing the relative energies of two rivers flowing five miles an hour, one having

---

* Ross: The Diseases of the Nervous System, p. 99.

a width of a mile, and the other of a few feet. The intensity of both is the same, but the quantity of the wider is the greater. Lessen the current rapidity in the first instance to a few feet per hour and increase the rapidity of current in the small stream to fifty miles per hour, and the intensity of the smaller stream exceeds that of the larger, while quantities may remain alike.

Haskins * defines electro-motive force as the power which a cell or battery possesses of causing a transfer of its current from one place to another. For our purposes this may be termed energy. It is to current what pressure is to steam.

This energy or pressure when impeded or resisted lessens the quantity of the current. $C = \frac{E}{R}$ may be substituted.

The quantity of current over a nerve may, from this analogy, be regulated by the resistance of the nerve.

Greater muscular action calling for greater neural energy, the larger the sum of the combined cross-sections of the axis cylinders, the greater will be the quantity, the less will be the resistance, and as $C = \frac{E}{R}$ *the lessened resistance of the larger axis cylinder admits of a greater quantity of molecular movement for the same energy or pressure (rapidity of vibration) than when the axis cylinder is small.*

The relative total current force may rise and fall with change of plane, but, *cæteris paribus*, the relatively larger-sized motor nerve of the same level offers a better avenue for quantity than the sensory nerve. The tensions of differently sized wires may be paralleled by differently sized nerves in apposition. With the greater resistance of the smaller-sized sensory nerve, the tension would be great, and the current would be extremely easy thence to the larger motor fibers.

* The Galvanometer and Its Uses, p. 8.

When the sum of the centrally produced vibrations are suffi-cient in energy or pressure to overcome the resistance of a motor nerve, the precedent motion, characteristic of the muscle irritated, is produced in that nerve.

Resistances vary as cross-sections and numbers of nerves, hence, different motor nerves require different quantities of molecular energy.

By all means, the reader should avoid conceiving of the forces or energies, herein alluded to, as material entities. No force is a *substance,* but all forces are *conditions* or states of substances. When motions occur in masses or molecules, there has been a transferral, a propagation, of the motion from one molecule to the neighboring molecules. As there is no effect without an antecedent cause, the ultimate primary orig-ination of motion is the cause of all activity in the Universe, and all thoughts, speech, or other deeds, are the consequences of an endless chain of previous activities in the individual and his ancestors.

The potential energy, the possibility of moving, exists in muscles and await the stimulus to evoke contractions in them. Nerves are potentially energetic, and molecular movements which afford stimuli to the muscles, exist in motor nerves. These become kinetic, or active, when stimulated in turn by the neuroglia movements, which originate through irritations from other parts.

According to Gerlach,* and as can be seen by direct inspec-tion, the more the motor function of a fibril becomes evident, the larger is it relatively to associated evidently sensory fibrils in the same tracts or areas. Though many motor are smaller than many sensory, and great ranges of sizes exist between

---

* The Spinal Cord, Stricker's Histology.

fibers of similar function, associated action, I think I am safe in predicting, will be found to entail relatively larger fibrillar cross-section upon motor function.

Sensory fibers and cells are relatively smaller than motor fibers and cells when associated physiologically with one another.

As before mentioned, *use* determines the size of nerve and cell, hence, the large cells of Purkinje, in the cerebellum, and the great size of those in the auditory nuclei, though related to sensory function, must be taken in connection with the network of finest fibrils in the auditory nuclei, which Meynert mentions as "of all the masses of the gray floor the most closely interwoven with fine fasciculi."

Henle * mentions an average of fifteen mmm. for motor and ten mmm. for sensory diameters, but calls attention to the relativity of these measurements, the larger fibrils reducing to smaller in their terminations. Efference is quantitatively greater than afference, and, as Spencer notes, the centrifugal waves are the stronger.

When the requisite quantity of energy is produced in the central area, through afferent irritation, the reflex motor act follows.

The differences of stimuli rates required to contract red and pale striated muscles, tetanically, are ten in the former to twenty to thirty in the latter, per second, and indicate, probably, better facilities for quantitative conduction in the case of red muscles, though the rates of contraction of muscles being quicker in insects than in frogs, and in heart muscle than in intestinal smooth muscles, bear upon differences inherent in the muscles themselves. Sympathetic fiber resistance, however, is relatively greater than in other nerves, owing to the relatively smaller-sized fibres that system contains. The 19½ muscle beats of

* Handbuch der Systematischen Anatomie, Nervenlehre.

Helmholtz denotes the meeting of quantity impulses needed to convulse a muscle.

Francke and Pitres* deduce from their experiments that motor excitations transmitted through the cord pass only 10 metres per second.   Helmholz and Baxt say that cooling a motor nerve caused the speed of transmission to fall from 33 to 30 m., and when heated, accelerated it 89.4 m.   According to Foster, sensory impulses vary from 26 m. to 44 m. or more a second.   He concludes that sensory impulses do not differ in speed from motor, essentially.

As between sensory and motor speeds, I incline to think that relativity here comes in with resistance and energy, proportioning the speed regardless of mere name of the nerve.

The diffusion caused by fright causing the knees to knock, extreme weakness, and the fact that over-exercise of one set of nerves occasions weakening of the general nervous system, shows the dependence of the entire system upon the relativity of its motions, and that special abstraction of energy, either in intensity or quantity, entails general abstraction of nerve force.

To account for the great number of nerve fibrils, we must restrict the wave lengths to sets of nerves especially adapted to entertain wave lengths, or consider that nerves differ in susceptibility to different intensities or quantities of molecular motion.   One wire suffices in telephony and telegraphy to pass countless millions of vibrations of different lengths and amplitudes.

By analogy I would prefer the view that *all nerves convey all ranges of wave lengths common to any nerve, but the cross-section determines the quantities of motion they were capable of conveying.*

* Gazette des Hopitaux, Dec., 1877.

Simple reflexes occur when the accumulated centric vibrations are generated in quantity sufficient to saturate the motor nerve or reach the point requisite to induce the production of the requisite quantity.

Compound reflexes are produced through the quantity sufficing to supply or propagate force in another nerve.

The alternate exertion of bilateral groups of muscles consists of the easier transfer of irritations from the exhausted side to the nerve roots of the rested side.

The walking of vertebrates generally, and the swimming of the fish through alternate exercise of muscles on both sides are instances.

Compound reflexes are only coördinated through much training, and involve not only the building up of the necessary correlating fibers, but the regulation of their sizes to the demands made upon them.

In what does inhibition consist? We know that in spite of stimulation, a muscle may resist action. A channel for diffusion seems to be afforded, but two opposing currents, acting simultaneously, nullify the effects of both. When two contrary impulses are generated, the resultant is zero. In a checked impulse to go forwards, for example, the impulse necessary to do so is too feeble, the generated waves are not sufficient, but in doubt whether to turn to the right or left, both groups of motions are alternately feebly excited, but strongly enough to swerve the body in both directions, but other channels are also operated upon, diverting the currents. In complex phenomena such as these we are apt to lose sight of the simple conditions upon which they depend. One reflex being possible, another is also possible, the correlation of these two reflexes succeeds by division of currents between them, and the possible number of compound coördinate reflexes which may follow are incalculable.

Putting this into nerve-force terms, we have : general sensibility 10^7 to $10^7 \times 2$ sufficing to saturate a motor nerve, the
cross-section of which admits of $10^7 \times 1\frac{1}{2}$ as its quantity, the
simple reflex results. The remainder of the stimulus being
diffused in adjacent gray substance, but if added to by a subsequent reverberation of $10^7 \times 2 = 10^7 \times 2\frac{1}{2}$, then a more distant organ in addition may be stimulated, their synchronous
action requiring the generation of a higher molecular action or
the successive discharges from center to both channels. These
motions are only practicable when the nervous system has
undergone adaptation, adjustment to the conditions necessary
to provoke them.

A protozoon which adapts itself to a change in environment
can only do so through environment reacting upon its molecular structure and making such rearrangement of its sarcodal
elements as will render further life possible to it, just as continued abrasion and impacts on palms of hands and soles of
feet harden the dermis mechanically, so that inconvenience
from labor or much walking is not experienced as formerly, so
in nerve and muscle tissue the physical basis of "becoming
accustomed" to anything is in a mechanical or molecular re-
arrangement of one or both of these organs. Certainly the
ability to lift heavy and heavier loads is accompanied by and
dependent upon the growth of muscles and, inferentially,
nerves.

Reflexes consist of stimuli evoking waves in such speeds,
lengths, and quantities as comport best with molecular motor
movements sufficient to excite muscles. The cord reacts best
between nerve elements of the same level. For reflexes of
different levels the stimulus must be increased or be produced
by summation of other molecular movements to the intensity
and quantity that will do the work. When things are done in-

stinctively, it is by virtue of the readjustment which makes the reflex easy. Habit, instinct, adapt means to ends so effectually in proportioning widths of axis cylinders to lengths, and the whole resistance is lowered to the point which makes reaction to constantly recurring certain stimuli so exact and inevitable as to make deviation from acquired facility of muscular motion extremely difficult, compelling the painter to use his left hand occasionally, to overcome the precision of touch which constitutes his "style."

CHAPTER VII.

PHYSICS OF THE SYMPATHETIC NERVOUS SYSTEM.

With regard to "trophic nerves," the general belief is gaining ground that none such exist, but that nutritive changes which follow nervous lesions are referable to vascular disturbances. This is so satisfactory that it at once transfers the mysterious and inscrutable to simple and knowable processes. Trophic influences are placed within vaso-motor precincts.

Lawaschew,[*] of St. Petersburg, under Professor Botkin, elaborates the subject, and holds that dilatation of vessels consequent on nerve lesions are often precursors of "trophic changes" in the tissues and due to "irritation of the vasodilator nerves," and not to paralysis of the vaso-motors.

The foremost vaso-motor stimulant is temperature difference, but other modes of molecular motion influence that system either directly or indirectly.

The vascular function is a highly nutritive one, and where first one part of the body and then another demanded blood supply in excess of other parts, just such slow rhythmic pulsatile motions as are observed in the vascular system would follow engorgement and dilatation of one part would passively effect distention of distant parts subjected to regurgitant pressure, a to-and-fro irritation of the arterial circular muscles would arrange commissural tracts facilitating transfer of equilibration and maintenance of tonus. Dilatation of one part stimulating molecular diffusion from other parts to act as constrictors, the calibers of the vessels are kept within limits.

---

[*] Centralblatt für d. Med. Wis., 1883, p. 193.

An equilibrating electrical pumping apparatus could be constructed upon this principle. Around a flexible tube pass segments of electro-magnetic clamps, so arranged that when one set of clamps are pressed apart by over distention of a part of the tube, it "short circuits" the current passing over a wire which extends the length of the tube, and causes the distributed current to act energetically in the part needing constriction, entailing an expenditure of force which drops the general normal several galvanometric degrees.

Distention acts as a generator of force (the increased heat and other molecular interchange is evident in distended parts); this suffices to raise the nerve vibrations in that part, and with increased blood consumption the facilities for both muscle and nerve motion are locally increased, but the diffused motions pass to the general sympathetic system which, having its plane of action and exerting its usual tonicity force elsewhere, distributes the pressure. It is evident that this can be effected but slowly, the mechanism for quick equilibration between all parts not existing, but as Carpenter * says : "When once a mode of nutrition has been fully established it tends to perpetuate itself," and, he might have added, differentiate and develop. Between points remote or contiguous wherein fluctuations are extreme, there would arise better facilities for equalization through simple development by extra use of the fundamental system.

Antipodal head and abdomen, with their complementary alternate excess and diminution of blood, have built up the splanchnic and cervical with the spinal sympathetic commissures. With dilatation of one pole a channel affording least resistant passage for a constricting influence would be a desideratum, and forces acting on that line would build it up.

* Human Physiology, p. 556.

Granting the constrictor abilities of the splanchnic and cervical, how shall we account for the dilatation through stimulation of nervi erigentes, chorda tympani, and muscular nerves?

Dilatation is a nutrient reflex. Upon muscular excitation there is a reparative demand. If a muscle be stimulated to contraction, it would seem proper that means of repairing the waste involved in the contraction should be provided. The osmotic power of the vessels is drawn upon with muscular action, occasioning a secondarily induced rush of blood to the part being nourished. This being the established action of the entire nerve distribution in the parts under consideration, that mode of action would adjust both cerebro-spinal and vaso-motor systems to its repetitions, and it must be borne in mind that vaso-dilator fibers run chiefly in the cerebro-spinal, vaso con-constrictor in the sympathetic nerves, but even admitting the dictum of Dastre and Morat.* that vaso-dilator fibers exist in the vago-sympathetic trunk, concomitant phenomena, such as dilatation following exercise of a part, could develop the molec-ular susceptibilities of a nerve constantly associated with defi-nite workings of other nerves as to render the dilatation constant, whether the motor or the sympathetic branch were irritated.

Associated serviceable habit developes the most extraordinary capabilities throughout organic life, especially evident in the emotions, as Darwin has shown so well.

Bernstein † concludes that "carbonic acid dyspnoea" is chiefly expiratory in character; the dyspnoea caused by want of oxygen is chiefly inspiratory. These relative influences upon the respiratory center of excess of CO_2 and want of oxygen in the blood can only be demonstrated by cutting off the vagus control.

---

* Arch. d. Physiol. Norm. et Path., 1882, 1, 2, and 3 ; and Centralblatt f. d. Med. Wis., 1882, p. 731.

† Du Bois-Reymond's Archiv., 1882, p. 313.

Blood condition in a center can become associated with distant movements, calculated to alter the nutrition of the irritated point for the better, and whether through paralysis of a strand or through stimulation, the adjustment of the nerve to its function may result in diametrically opposite modes of working. The apparently erratic conversion of constrictors into dilators and *vice versa* are best accounted for by Lepine's* experiments on the frog's sciatic, the conclusion from which is that the same fiber may act as dilator or constrictor, *according to the condition of the peripheral mechanism.* Foster † suggests that the paths along which the impulses of afferent or central origin issue as efferent impulses are determined in part by the condition of the cord and character of the afferent impulses, or of the central disturbance.

The vaso-motor phenomena attending *every* reflex, whether cerebro-spinal or sympathetic, are of more importance than could be judged from their little consideration by writers. The assumption of a vaso-motor accompaniment to every act of the nervous system involves an explanation of much hitherto considered inexplicable. No blood, no action; plus blood, plus action. Every vital act incurs expenditure, sensation as well as motion. The irritation of every cell causes waste, and necessitates repair. If the reparative (vaso-motor) power be annulled, the function ceases with death of the part. Regeneration of nerve areas must be regarded as identical with that of any other tissue. The brain is nourished precisely as are other organs, reflexly through irritation at the point needing restoration. Molecular change, depletion, expenditure is the irritant. The amœba moves more rapidly when hungry than after being fed. The vaso-motor is the reparative system, and

---

* Comp. Rend. Soc. Biol., March 4th, 1876.
† Physiology, p. 284.

its reflexes must be considered with reference to that function, regardless of but little else. Such visceral parts as use blood most rapidly are in the greatest condition of vaso-motor irritation. Habit fixes the quantity of blood and regulates its afflux. The vaso-motor connections with the cerebro-spinal system have the same significance as their connection with other organs. Splenetic, gastric, uterine, hepatic, etc., irritation, through want of blood, starts the reflux, and no matter whether sensory or motor nerve needs restoration, the irritation of either will start vaso-motor action as readily as to abdominal viscera.

The connection of sympathetic fibers with both sensory and motor cerebro-spinal nerves has no other value than that the vaso-motor system recks of molecular motion without discrimination as to whether those motions produce sensations or muscular actions. If a sensation pass over a spinal nerve, molecular motions are set up in that nerve and stimulate vaso-motor action ; if a motor nerve be engaged, both nerve and muscle thereby excite vaso-motor afflux of blood to them. But due regard must be paid to the blood conveyance in the entire animal, for a disturbance at a distance may render stimulation of other parts inoperative. Hence, the variable nature of sympathetic workings may be represented by the active points ABCDE, having a tonus of 10 along their vaso-motor filaments. A, being stimulated to contractile effect 20, BCDE vessels at first are dilated by blood pressure, but by diffusion the general tonus is distributed as 12 to all points. Let E demand repair through exhaustion, lowering the tonus of E to zero, then the remaining tonus of all falls to 9.6. Now let C and E require blood before the vessels at E have been restored to their usual calibre, ABD will be in tone, C dilated, and E extremely dilated. It is plain that a strand ordinarily acting as a constrictor at A, could not thus act at this stage, but would appear to act as a dilator, through its stimulation having aided in raising the tone of C and E, the general distribution causing the tone at A to fall.

The vascular tonus being normal, generally, in the " indifferent regions," there will follow dilatation if other vascular regions are dilated, through the diffusion acting first to re-establish tonus, the dilatation being compensatory in the region apparently stimulated.

Vaso-constriction is the usual effect of vaso-motor irritation. Vaso-dilatation may become the normal effect of stimulation of a vaso-motor nerve, by associated influence of cerebro-spinal nerves, or by diffusion of vaso-constriction to adjacent vascular regions with minus tonus, such minus tonus may be the effect of the same stimulation which produces the apparent dilatation, through action stimulated in those adjacent parts (muscular, usually), or the minus tonus may be caused by synchronously operative distant causes. Habit, adaptation, or readjustment will thus implant dilating instead of constricting function upon certain vaso-motor regions. In the simplest possible language, if other accessible parts need constriction or restoration of tonus most, diffusion in lines of least resistance will occur, and the apparent dilatation of the near point is the afflux of equilibration. A part having this mode of action established would develop facilities for diffusion of stimulation to the other parts and, by relativity of forces, the near part would be paralyzed, and dilatation become extreme in that part on stimulation.

CHAPTER VIII.

PHYSICS OF THE NERVE CELLS.

Thus far I have purposely avoided any mention of nerve cells or ganglia participating in nervous or mental workings. When I affirm that there are many and good reasons for regarding the main function of the nerve-cell to be *histogenesis*, and that all the workings of nerves proper, and the sensitive basis substance, such as the neuroglia of the spinal cord and brain cortex, are best, or equally as well, studied without reference to the nerve cells, I state that which needs defence; but the acceptance or rejection of the histogenetic function view and the refusal to concur in or adherence to the olden view of the nerve cell being an energy-producing body, does not affect, in the least, the value of the deductions or inferences up to this point, nor those to follow. *There is no dependence of the inferences mentioned upon the correctness or incorrectness of my theory that the nerve cell is a cell, correctly so named, with highly reproductive powers, and very small and few other powers.*

In saying that this assertion needs defence, it must be remembered that the olden theory of force-production function on the part of the nerve-cell needs equal defence, and is an assumption of the broadest kind, to sustain which the most trivial reasons have been offered in the face of much evidence to sustain a contrary view.

March 11, 1881, I advanced my histogenetic nerve-cell theory in a paper read before the Illinois State Microscopical Society,* and made it the subject of my American Neurological

---

* An abstract of which, full of typographical errors, appears in the Chicago Medical Review, March 20th, 1881.

Association thesis at its June, 1881, meeting. The grounds for my conclusions are as follows:

The nerve-fiber and not the nerve-cell is the first to arise in forms above the protozoa, as in Kleinenberg's Hydra, Pseudonematon, and other low worms. In noto-chordal animals such as the amphioxus *an elaborate system of nerves exists without a nerve-cell being present.*

Free nuclei and neuroglia abound, such as Schmidt[*] and others[†] observed in the human embryonal nervous system, where nerve-fibrillæ cluster about nuclei, the fibrillæ arranging themselves in rows and fasciculi. "The pia mater extends from the brain and spinal marrow over the peripheral nerves in the form of a single sheath. The whole nerve consists therefore, like the nerve of an insect, only of a bundle of granular fibrillæ, which, in pursuing a wave-like course, are placed parallel to each other, and surrounded by their sheath and neurilemma" (Schmidt, op. cit.) He also mentions "mother-cells" packed with larger and smaller nuclei. Multiplication of nuclei by endogenous modes ceases at one stage and gemmation begins.

The nerve-cell is born of the original protoplasmic cell, and is highly productive of granules, which form axis-cylinder processes.

Deitl [‡] agrees with Mayer in assigning axis cylinder reinforcing properties to nerve-cells, and both consider the sympathetic nerve-cell as a reserve material for the reproduction of the nerve-fibers that have become unsuitable for the transmission of nervous influence. "For this development the nerve-cells utilize the elements of the blood, notably the red globules. We find in the sympathetic in the vicinity of blood-vessels special elements formed of a fundamental substance and numerous nuclei, analogous as regards form and micro-chemical reactions to the blood globules. The fundamental sub-

---

[*] The Development of the Tissues of the Human Embryo, Journal of Nervous and Mental Disease, July, 1877.

[†] Histology of New-Born Cortex.   Brain, July, 1883, p. 287.

[‡] Sitzungb. der K. Akad. der Wissensch., 1874.

stance, closing gradually around these nuclei, form masses of protoplasm which slowly take on the appearance of nerve-cells. These, apolar at first, later present a prolongation which, uniting itself with one or another cell, forms a nerve-fiber." Mayer's researches on the sympathetic, and Deitl's upon other ganglia, make the application general. Kölliker, Beale, and Henle hold the division of nerve-cells into others. Thus not only is the reproductive faculty highly exalted in the direction of axis-cylinder formations, but also for cells of its own kind. It is a simple biological principle, that development of a faculty in one direction involves suppression without necessary extinction of other attributes. It is inconceivable that the nerve-cell should possess two highly differentiated peculiarities. Hence its energy origination must be in abeyance.

Fibers may be and are produced primitively independently of nerve-cells.

The brain and spinal cord of the embryo is one large nerve-cell.

The nerve-cell is not necessary to nerve action, for in the lowest vertebrate there is an utter absence of such cells, the fibers alone affording passage for this form of molecular motion.

Dr. Roler, of Chicago, obstetrician, told me five years ago of a case in his experience of an anencephalous child, whose vigorous kicks in the course of delivery led the doctor to diagnose anencephalism before birth, as he had just read in the *American Journal of Obstetrics* of a similar case of explosive reflex irritability in the delivery of a similar monster. Now, if this is a common anencephalic accompaniment on the basis of nerve-cell genesis of force, how can this be accounted for when the millions of cerebral nerve-cells are absent?

The neuroglia sizes decreasing in the cortex in proportion to their development of fibrillæ (Stilling and Meynert regard the neuroglia as essentially a felt-like substance composed of the most indirect and finest nerve processes), show that sensation and molecular energy, of which the former is one expression of the latter, reside in the neuroglia, and that fibrillar development is the great desideratum in increasing facilities for molecular interchange of a nervous nature. Every cell in the body, every nerve-fiber is capable of molecular motion, and though the nerve-cell is, as Spitzka * says, " the ultimate functional center of the nervous system in man and other higher animals," the parallel fact, which he also cites, that increasing intelligence and coördinations go hand in hand with increase of commissural nerve-fiber establishment, show that the increase in fibrillar connections requires the increase in histogenetic arrangements, such as we find eminently provided for by nerve-cell increase.

I regard the olden assumption of autocracy for nerve-cells as the last foothold of that superstition which pictured fleshless bones stalking abroad at midnight. The sizes of these cells would be in proportion to the stimulation they sustained toward granule parturition, and the immense number of fine fibrillæ near the internal auditory nucleus is a case in point. Cells such as the auditory, and the giant pyramidal of Betz, are certainly under constant strain. When any mechanical, chemical, or electrical irritation suffices to stimulate nerves, the creation of a special stimulating apparatus other than such as exist in amphioxus, dipnoi, etc., would be unnecessary. Goltz tacitly admits the direct stimulation of the nerve-fibers without cell intermediation, in criticising Ferrier's experiments.

The associated atrophy of nerve-cells and fibers, such as Davida† found in perobrachia (arrested development of the forearm) and pathologists frequently find in tract degeneration, bear in the direction of the failure of cell stimulation through nerve destruction, for an organ ceasing to act through want of stimulation will atrophy, but an organ which stimulates need not die because that which it stimulates has perished. The nerve being the stimulator and the cell the stimulated body, nerve-cells may atrophy without entailing destruction of nerve-fibers, except where the so-called trophic relationship exists,

---

* Architecture and Mechanism of the Brain.
†Virchow's Archiv, April, 1882.

which seems to indicate the dependence of the nerve upon the cell for its elements; otherwise Wallerian degeneration of nerves would be centripetal instead of centrifugal. The nerve destruction in this case does not cause cell destruction, but if the stimulating area in which that cell reposed were function- less, the cell would die, *but not the centripetal nerves.*

The diversified shapes of nerve-cells seem to bear relation to the composition of forces rather than diversity of function. The lines in which impinging energies act determine shapes of other tissues, and speculations in this direction applied to nerve-cells would be fruitful. The fusiform bipolar cell could result from forces traversing it; the globular unipolar, from forces terminating in it; the multipolar, from varying quanti- ties of energy acting in several directions; the pyramidal, from having its point of greatest tension at the base, with least re- sistance at the large tapering process. Always remembering that these cell activities appear to be mainly histogenetic, and the stimulation it receives serves to elongate the process and promote the more rapid formation of nerve-granules and fibrils, *but there is nothing to prevent the cell-contents from acting pre- cisely as do other fibrillæ in the conduction of motion.* The gan- glionic bodies Ranvier, Max Shultze, Schmidt,* Deiters, and others have shown contain multitudes of fibrillæ and granules of minute sizes.

All investigation goes to show the precedence of the nerve granules which are formed into rows and fasciculi, and that nerve action precedes the appearance of the nerve-cell. The lat- ter is formed at a stage when greater multiplication of granules and fibrillæ proceed, and the cell is formed *from* these pre-ex- isting elements, in my opinion, for the purpose of hastening the granular and fibrillar creation.

* Schmidt's researches and engravings in the Journal of Nervous and Men- tal Disease, Jan., 1879, are excellent.

11

CHAPTER IX.

PHYSICS AND CHEMISTRY OF THE NERVOUS SYSTEM.

However the Liebig, Meyer, Voit, Traube, Matteucci, Heidenhain, Hermann controversy may be reconciled, an extension of the hypothesis of the latter to the protagon activities of the brain affords satisfactory analogies. The phosphorized nitrogenous cerebral substance exhales the same gas, CO_2, into the blood, and uses up oxygen as does muscle. Cholesterin bears some such relationship to the nerves, urea and other such crystals do to the muscles, and in the elimination of cholesterin the liver is associated with the nervous system as the kidneys in the elimination of urea and its congeners are with the muscles. The distinction between cholesterin and the urea group seems to be the absence of nitrogen from the former and its presence in the latter. The distinction between protagon and inosine is the presence of phosphorus in the former. The oxidation of the phosphorus and its union with the cyanities, with liberation of cholesterin and CO_2, with accompanying propulsion of exploded particles apart, the seggregative nerve property, have complementary characteristics in the muscular ammonium cyanogen change to urea and concomitant conversions producing contractions by aggregation of molecules. Such increases and decreases of bulk and specific gravities in the union of chemical solutions are familiar to tyros in chemistry.

Phosphorus has a great affinity for oxygen, and cyanogen enters into combination with the non-metallic elements, as though it were itself an element. Chlorides, bromides, and iodides of cyanogen are easily formed, facts which have both physiological and therapeutical bearings.

Kopp * says phosphorus expands 3.43 per cent. at the moment of fusion at 44° C., but contracts like wax and sulphur on cooling. Phosphorus assumes several different forms under the influence of causes apparently trifling. Thus, while in muscles there is combustion and shrinkage, in nerves there is combustion and explosive repulsion.

The increased cerebral blood-supply during thought and the intoxication produced by oxygen are evidences of the rapidity of protagan interchanges.

The evolution of heat being definite in oxygenation, the more rapid molecular interchange in higher animals is accompanied with higher heat-rate. Electricity is definitely produced also, as shown by Hermann in his inogene studies, and where this had served a useful purpose in perpetuating the life of an animal with a morbid condition of muscular excitability, it is likely enough that natural selection would seize upon it. The heat given out during the combustion of a compound body is less than that emitted by the combustion separately of a quantity of each of its constituents equal in amount to that present in the compound burnt.† The amount of heat disengaged by H, CO_2, CHN, Fe, Sn, SnO, and P with equal weights of oxygen is nearly the same, except that in phosphorus it is the greatest.

The heat-rate of an animal is at first only an epiphenomenon, but, like everything else which occurs, must be reckoned as in its environment, initially a product of life, it must be allowed for when it increases the molecular speeds by reacting

* Liebig's Anal., xciii., 129.
† Miller's Chemistry, vol. i. p. 401.

upon the basis substances. The heat evolved by the animal adjusts the molecular rates, and becomes a part of the life-phenomena, alterations of which entail changes proportionally more or less formidable. Similarly adaptation or readjustment suggests that, in muscular susceptibilities, various stimuli being adopted as a normal, it is conceivable that muscles, as well as nerves, must be considered as educable or adjustable, and become non-responsive to stimuli often repeated through molecular rearrangements. *Muscles hence must vary between themselves as to their selective response to stimuli,* and thus much of the burden of compound reflexes may be greatly lifted from the nervous system.

The protozoon exists in an environment of nitrogen and oxygen, the former inert and the latter a food in its affinities for protoplasmic derivatives. The combined H_2O have a main significance as a vehicle for molecular interchange, facilitating the necessary atomic and molecular translocations. Nitrogen has a persistent tendency towards its free inert state, and this very disposition confers upon it great physiological importance. On the other hand, oxygen has a great antipathy to uncombined existence. These two mechanically mixed ingredients of the air play in biological phenomena complementary parts.

Phosphorus in its assimilable form of the tricalcic diphosphate conveys its bone-producing properties in connection with the attributes of gelatin for its solution. The sudden appearance of large-boned vertebrates in a phosphorite epoch was noted by Prof. E. D. Cope,[*] and Filhol[†] subsequently remarked the same thing. The profound modifications which the palæontologists convince us animal life underwent through

---

[*] Hayden's Reports.
[†] Phosphorites of Quercy, 1878.

accidental advantageous surroundings undoubtedly include the presence of assimilable phosphorus in plants, and accompanying physiological possibilities for its assimilation. The differences in the excreta of the vertebrate division show that materials made vital to a higher life through systemic absorption and utilization are inert excrementitious matter in the lower animals. The acidity of muscle and nerve substance in connection with blood alkalinity, render possible the conveyance of alkaloids, and make it likely that soluble alkaloidal hydrocarbons of the neurotic group, assimilable by the organism, have sufficiently close molecular resemblances to the acid protagon as to account for their mutual affinities and biochemistry.

The differences in the physical attributes of the great neurotics are not apparent in their respective proportions of carbon, hydrogen, nitrogen, and oxygen. There are some marked relationships between quantivalence, atomic weights, numbers of molecules, and physical properties of compounds, as well as in elements; but, aside from such examples as Kopp's law of boiling points of ethers, the temperature increasing with the numbers of CH_2 atoms; in general, the chemical formulæ of substances afford no clues to their *modus operandi* with respect to heat, light, electricity, etc., and comparatively little has been done to tabulate from direct experiments such peculiarities. The rationale of the neurotics and medicines generally is a a tempting field for speculation. The bitterness of most tonics, the fact that tannic acid kills paramecia, the alkalinity of ammonia in connection with the acidity of nerves and muscles, the poisonous properties of phosphorus being increased by fatty matters, the nutritive interferences in toxæmia, etc., with such investigations as Spitzka* on strychnia would afford the

---

* The Anatomical and Physiological Effects of Strychnia on the Brain, Spinal Cord, and Nerves, Journal of Nervous and Mental Disease, April, 1879.

chemist most satisfactory considerations. The carbonization of the blood in hydrocyanic acid poisoning refers to a central toxic action. Hensinger, quoted by Darwin,* notes that white sheep and pigs are differently affected by vegetable poisons from colored individuals. Such differences may be due to the presence or absence of pigment compounds which have affinities for or resist the influence of certain poisons. In the constitution of protoplasm, as well as that of any organic or inorganic compound whatever, there is an ever-present necessity for the *absence* of certain molecular groupings which would destroy the combinations under consideration or cause them to act erratically. The cell environment is reached by adaptability, and in the differentiation of cells it is easily seen that what would be nutrient to one may easily poison another by *a*, combustion conversion complete; *b*, lesser affinities existing between toxicant and the molecules.

The life of the cell depends upon the absence of these deleterious molecules for which there are affinities precisely as there is necessity for somatic absence from fire. Prussic acid presents the simplest example. The nitrogen therein is in a dangerously assimilable form, and its sudden surcharging of the nerve-centers with carbonized blood paralyzes the body.

Herbert Spencer † elaborates the anæsthetics' modes of action. No philosophy can explain nerve action by excluding nerve and nerve-center therapeutics. As certain drugs have specific affinities for certain groups of nerves, it follows that the times of action of such groups differ from others, and in such selective affinities we have indications of the differential molecular workings of both nerves and neurotics. If the cell becomes adapted to a certain environment, and the genus of a

---

* Origin of Species, p. 18.

† Principles of Psychology.

certain chemical combination be poisonous, one species of such genus may through adaptation cease to be a poison within relatively restricted limits. A readjustment of the entire body may occur permanently for the race or for the individual on the basis of toleration. For instance, phosphorus has been adopted as a normal ingredient of cerebral tissue; but antimony and arsenic in the same group remain as poisonous as phosphorus is to those animals whose assimilative adjustment for it has not reached the stage it has in man.

On the principle that a millionth of an inch is as much of an entity as is a million miles, molecular distances, though unseen and unmeasurable, inferentially exist. The vast range of tissue solidities throughout the body, the coarseness of the deltoid muscle, and structural fineness of other muscles, the varying permeability of organs and their ability or inability to resist molar or molecular motions of graded degrees, the adjusted calibers of vessels to their contents, of cell and cell granules aggregation in dermis, epidermis, lung, liver, serous membrane, and bone, like the callous palm of the laborer, or the bicipital growth of the oarsman, indicate the adjustment of molecular distances to the exercise they undergo. In iron wire the passage of so imponderable an influence as electricity in one direction adjusts the atomic distances in lines of least resistance. This is so well known to electricians that telegraph companies often order their current directions reversed to habituate the flow to indifferent directions. Analogy forbids any other view than that nerve tissue is no exception to this law of adjustability, and that constant causes produce constant effects in creating densities of nerve structure consonant with the work performed. The softness of the "*portio mollis,*" the hardness of the other nerves, notably the optic, are significant and seem related to the differences in wave length conveyance. Roughly estimating the comparative densities of the auditory and optic as 1 to 10, the former related to vibrations of from 40 to 4×10^4, the approximate equation is obtained $\frac{757 \times 10^{12}}{4 \times 10^4} = 18925 \times 10^6$ for the retardation which the highest light ray, violet, must undergo in comparison to that of the highest note that is heard. The assumed greater density 10 of the optic nerve dropping the ratio only one power $= 18925 \times 10^5$, and even were the optic one hundred times denser, the whole differences would not suffice to account for the transformation undergone except by calling in the peripheral apparatus and its rhodopsin. Nevertheless it is evident that the centripetal differences create some changes in densities.

Stieda calls attention to the axis cylinders of the second root of the auditory nerve as thicker than that of any other, the first having delicate fibrils. The second has a ganglion upon it like the intervertebral. What these changes in destiny measure, and the precise molar and molecular arrangements may be, are, at present, only conjecturable, but their existence is beyond dispute.

The celebrated Chladni figures, made by the action of musical vibrations upon sand arranged on glass plates or membranes, as yet have not had the causes of their appearances being related to their corresponding notes discovered, but the rule is, that *to a determinate note belongs a determinate figure for the same membrane or plate, and that a figure is more complicated as the note producing it is higher.*

The only observable aspect of these figures in the nerves is density, and here we must not mistake gross appearances for intrinsic differences, for connective tissue increase may sclerose. I refer to the densities of the intimate proper nerve substance which cannot always be inferred from simple inspection. The presence or absence of water, fatty matter, protoplasmic soft substance, etc., with the distances apart of granules and the intimate density of the granules themselves, present a complex upon which impinging forces would act to arrange, mould, adjust, and rearrange.

Now, the persistency of a force, its frequent recurrence, could not do otherwise than produce a definite arrangement, intermolecular, intergranular, and with regard to extrinsic but associated elements. We are forced to speak of the *arrangement* of a nerve as including its density and the other factors mentioned.

A telephone and its wire used for years in a mine, to acquaint the engineer above ground with the workings of the valves in the pumps below, was found to have so adapted itself to the almost uniform vibrations as to give other sounds imperfectly.

A nerve which has undergone adjustment acts reflexly *instinctively*, and where not only in the individual but in the species, impressions constantly recur, the nerve arrangement would be inherited. Impressions common through long periods to all animals would arrange common nerves in common manners. Impressions only experienced by a species or race would differentiate nerve structure therein above the common, provided the recurrence of the impressions were frequent.

Impressions only experienced by an individual result in a reintegration of the nervous system at the expense of disintegration of racial and species adjustment, and such breaking up of inherited arrangements we know to be exceedingly difficult to bring about.

The more perfect adjustment of nerve elements to the transmission of vibrations constitute instinct, and I extend this facility of motion to sensation, feeling, motor muscular activity, and to all mental processes.

CHAPTER X.

MENTAL PHYSICS.

Reason or deliberation, or in the spinal nerve action, hesitancy in reflex response is indicative of resistance to the passage of a new set of impressions, the inharmony of the adjustment to the influence causing the deliberation. As soon as the vibratile term most in accord with the impression has been determined upon, the action which follows is the effect of most accordant routes having been selected.*

Hammond's dictum, that where there is gray matter there is mind, could be extended to nerves, thence to muscles, thence to protoplasm. Mind is a thing of such degrees that from its evolution in man to its beginning differentiation in the protozoon we can no more demarcate the steps of its growth than we can tell where the seed ceases to be a seed and becomes a tree. Huxley's comparison of the mind to the horologity of the clock is a just one. Stop the pendulum, or derange the mechanism otherwise, and the time-keeping capacity of the clock is in abeyance. Interference with any of the living cells in the body, according to their states of mental integration and relationship, will derange the nervous and mental mechanism and its life-subserving workings, the mind.

Cope assigns mind to the protoplasmic cell; so do I, but in the sense of mind being only on a par with all the other chemical and mechanical motions of the body. It is in the

---

* Further elaborated in my "Contributions to Comparative Psychology," Science (N. Y.). "Instinct and Reason," May 28th, 1881. "Origin of Language," July 23d, 1881.

metazoa the coördinated activities of the associated cells, of every cell in the animal, regardless of development, shape, size, location, or degree of vitality. As Spencer claims, it is inseperable from other vital workings.

As in the amœba there cannot be movement without molecular consumption, nor repetition of this movement without re-supply throughout animal life, we see the ability of tissues to maintain their integrated arrangement and yet to be subject to waste and repair. This nutrient possibility of maintaining the adjustment, once fastened upon the organism, cannot be · denied to nerves; the gross muscular and other persistencies imply molecular arrangement persistence in the face of destruction. When in the organization of the metazoa definite organs as cilia appear, adaptation to repetition of movement is apparent and each cilium will, in repeating a movement, cause a repetition of associated molecular motions in its sarcode directly related to these motions. Now, if with the first repetition of that ciliary motion we grant that similar molecular movements to the initial motion were reproduced, then we have the necessary conception of *memory.* Let the ciliary motions and the accompanying molecular sarcodal changes continue indefinitely, memory becomes merged into hard and fast lines of energy capabilities, definite molecular translocations, and so far as an animal can or cannot be said to remember an act which is part of its existence and of momentary recurrence, it is conceivable that it would require a decided disturbance of its structure to cause it to forget. But as Ribot* points out, we remember by forgetting, and it is through rendering the reflex susceptibilities to recurring stimuli automatic and part and parcel of our structure that we are enabled to integrate this kind of memory with instinctive motion and cease to regard it till disturbance occurs.

* Les Maladies de la Memoire.

An associated molecular change which faintly disturbs the protoplasmic granules concerned in the ciliary motions will thus revive the memory of those movements. Let an internal fortuitous agitation of a sleeping hungry amœba take place, instantly the impression, the memory of eating is revived, through molecular changes induced identical with what occurs in eating, and may provoke appropriate grosser movements, such as protrusion of pseudopodia. This similarly, though with basal identity, occurs in the dreams of a hungry man. The irritations from the viscera produce a tension of the gastric and intestinal distribution of nerves requiring but the faintest correlative irritation from the brain to suggest the feast.

In the diaphragm to diaphragm telephonic transmissions of waves, the plexus of diaphragmatic vibrations oscillate in changed terms along the wire, to be revived at the other end. It is not necessary to have termini alike in the case of nerves, for whatever the initial oscillations may be, their translation into neural movements suffices to revive memory of similar movements, adjusted to in the cord and brain.

Conceive now of an associated apparatus, by which the telephone line, which we must regard as adjusted to definite waves, has its oscillations revived independently of the usual end apparatus. If these otherwise induced groups of waves be identical with those produced in the ordinary way, then the diaphragm vibrations will be as definite as though ordinarily induced. The phonograph traces its characters upon a metal sheet, and through them the sounds which produced them may be recalled. It matters nothing how these wire vibrations which make both character and sound are generated, if any other influence, similar to that which we know is at work upon nerve tissue, produces identical waves, however feebly, both or

either characters and sounds may be made. This is intended
to represent revivification of memory through association, such,
for instance, as occurs in recalling the odor of a rose when an
image of a rose is aroused in the mind.

The integrity of related organs, at least for a definite time,
is essential to memory. For example, the joints in arthropodal
members, and in metazoa generally, differentiate limb movements which adjust the muscular protoplasmic and nerve arrangements after repetition, and thus memory, in common with
other attributes of mind, is an osseous and a muscular as well as
a nervous phenomenon.

It is well known that convulsions so adjust the nervous system as to make their recurrence easier.

Delbœuf* adopts a view of molecular adjustment as constituting memory, but does not make applications such as are found
herein. Doubtless many other suggestions I have made herein
have been advocated before, but their synthetical arrangement
I shall claim, together with such matters as the nutrient
reflex, as original. It was at a recent date fashionable to ascribe
everthing to Aristotle. He was even accredited with suggesting
the rotundity of the earth, but it was not mentioned that he,
in the next breath, advanced reasons against such an idea.

Wundt, Spencer, and Bain are each close to the truth in their
several psychologies, but it is only by elimination of error and
adjusting with reference to extensive ranges of facts, that complete consistency will be evolved in such matters.

· To read of memory, as described by many writers, one
would suppose it infallibly reproduced facts and images; but
it does not, it is treacherous in the extreme, and only when
cultivated is it free from gross error. In hyperæsthetic states
it may reproduce minutiæ, nor is this to be wondered at when
we grant it ability to recall anything. In the main it is illusory,
and made up of all sorts of superimposed experiences. The

---

* Théorie Générale de la Sensibilité, p. 60.

limner takes advantage of the readiness of the memory to supply omitted points when he, with a few lines, leads you to imagine you have seen details he did not fill in his sketch.

Memory is an important intellectual omnibus, and, as Ribot says, consists of *memories*, which accounts for the loss of ability to recall certain things only. I would like to epitomize Ribot's excellent work in this connection, but want of space forbids. The inability to recall the letter F, in the case he cites, shows how memorized coördinations of muscles may be ineffectual through stimuli failing to act upon a defunct center for such coördination. Harrison Allen's recently invented "palate myograph"* would be valuable as an aid in determining the muscles involved, and inferentially afford knowledge of the coördinations of articulation.

The patient recorded by Dr. Ball,† who could not speak certain words he could not hear, had lost hearing memory of words, and could not pronounce them for want of the usual stimuli to such coördinations. Finkelburg's‡ suggestion of asymbolia for all phenomena of aphasia is a good one, as it generically classifies these failures of speech and gesture.

I think, since Exner's researches have given us better defined notions of a modified localization in the cerebrum, that the word registration (of which Spitzka is so fond) expresses the function of the adjustment or arrangement I have attempted to elaborate.

The registration is simply the facile possibility of revivification of the oscillations within the limits of nerve-arrangement and revivification of such registered impressions is a memory.

Memory in sensation has its analogue in the motor apparatus and motor incoördination, or erratic reflexes may be the

---

* American Naturalist, April, 1884, p. 38.

† Archives of Medicine, April, 1881.

‡ Berliner Klin. Woch., 1870, Nos. 37, 38.

congener of amnesia. Kussmaul* regards this motor memory as necessary to an explanation of aphasia.

Partial amnesia will be caused by disarrangements of certain tracts.

Temporary amnesia, their temporary failure of adjustment.

Periodic amnesia, also temporary, but recurrent, due, doubtless, to a nutritive failure.

Progressive amnesia, as in senility, the permanent involutional change which, being disintegrative, resists rearrangements, at least of a permanent character. The progress is from forgetting proper nouns to common, thence to adjectives and verbs, and finally emotional language and gesture, from the less to the better organized, from the complex to the simple.

Exaltation of memory would follow hyperæsthesia or raise of plane, and in some mania cases both the sensory and motor memories are much exalted.

Kussmaul's "Bewegungsbilder" suggests motor as well as sensory registration in the brain and nerves.

The brain, and particularly the gray matter, receives an enormous amount of blood. In no other part of the body is the nutritive function so active or so rapid. The stability of modifications and the dynamic associations between nerve elements can only be produced by nutrition (Ribot, op. cit.).

Between nutrition and retention there is a cause and effect relationship.

The slide-valve cut-off for steam cylinders is an arrangement in sequence, and its action may be likened to that motion of muscles which, reaching a certain point, is followed by related movements of other muscles. Certain acquired motor memories being invoked, are followed by associate memories for other motor apparatus. This acquisition is similar to the sensory association phenomenon of learning a thing by its con-

---

* Die Störungen der Sprache.

nection with some attendant thing or circumstance. One revived impression is apt to call up another. Does this require a
special association apparatus or a molecular motion relationship
of the sensory and motor events? We know that the first exists
and is developed *pari passu* with association powers, but how
is association effected? Having the machinery, how does it
act? The fibrillæ cross and re-cross each other, abutting in
gray centers and among nerve-cells without decided connections with each other or with the main strands. The resultants
of interference vibrations or musical beats are suggestive, but
insufficient, for beats cannot revive the notes from which they
were derived. The arrangements of both fibrils and nerves
suggest diffusion, and diffusion it seems to be, with the reservation that, when association by diffusion becomes definite between impressions or motions, tracts are built up through use.
The lines of least resistance are the lines associating memories
or other faculties, and in the cord those lines are in contiguous
levels, the network of Gerlach, and for higher and lower levels
the white columns of the cord. Diffusion having built up
these tracts, the sensations and motions are correlated by lines
which vibrate in unison with the tracts they connect, so that
forces from one tract may pass over the commissure and
arouse the vibrations of the other tract feebly or strongly.
Commissural fibers then are capable of conveying two or more
rates common to other fibers with which they are associated.
Coördinated movements must have fibers capable of being
simultaneously and successively vibrated through uniting
fibrils in which vibratory movements simultaneously and successively pass which are common to the main nerves stimulated. It is not necessary that each two nerves should have
a definite association system, notwithstanding the millions of
commissural fibers that exist.

It suffices alone that an impression should in the faintest degree resemble some previous impression to revive associated memory.

Given two simultaneous impressions, optic and auditory, the former abutting in the gyrus cuneus, and the latter in the supra-marginal gyrus, the coördination of these impressions would be effected by any preëxisting tracts. Such tracts, upon incessant stimulation, could undergo better definition, and this would seem to involve, not only the building-up of the callosum, but of definite cortical striæ formations, such as *a priori* one would expect to find relating the "visual" sense (in man the most important) with senses or motions the most frequently occurring.

It may not be necessary that the uniting fibrils should have the tension strength of the united fibrils. We may surmise, from the small work of network-fibril sizes, that it suffices if the uniting fibril acts at all, in which case the memory latent in each main fibril will be evoked or stimulated when the other two parts act. The most important thing to remember is that the *nutrient reflex* occurs with every cerebro-spinal irritation, the vaso-motor flash of supply to the points irritated. This seems to be the "something else," the *tertium quid*, for which psychologists have sought. The auditory and optic centers being stimulated, their respective blood-supplies are reflexly stimulated with an increasing disposition to nourish an intermediate tract, or to arrange more definitely the fibrils over which the association is made, the stimulation of one sense sufficing to produce faint vibrations of the other.

Motor memories are thus evoked, and both successive and simultaneous motor coördinations effected. The same plexus of fibrils answer for association of sensory and motor recurrences.

12

.The agreement of this hypothesis with the structural facts of the brain are, to me, very convincing.

"According to Meynert, the rôle of the cortex is not so much the liberation of motive force as the mediation of sensations of innervation (*Innervationsgefühle*). All determinable function areas which on physiological grounds must be considered to require a bond of association are found to present regional points of contact. Thus the symbolic field (for written and spoken symbols) approaches the fields of the upper extremity of the tongue and of hearing."[*]

Munk and Exner call attention to the tactile areas of the cortex being topographically identical with the motor centers of the corresponding periphery,

"*Perception reaches the maximum of clearness in attention*" well says Guiseppe Surgi.[†]

"We hear best in breathless attention," remarks Tuke.[‡]   A theory of perception thus would best be considered through analyzing attention.

According to my suggestion of nutrient reflexes, the increase of centric irritation would produce an afflux of blood to the part in functional activity.   The blood is there because of the work being done there, just as friction calls blood to the skin. Listening to a sound, the sight is not so acute.   Interferences between recollections or acute perception of impressions are less when they relate to different senses than when they are terms of one sense.   A landscape and piece of music may be enjoyed simultaneously, but not two sights or two kinds of auditory impressions.   In "breathless attention" the nutrient reflex not only affords one center an extra supply of blood for its activity, but denudes other centers, even the pneumogastric,

[*] Spitzka's Review of Exner's Pathol. Researches. Journal of Neurology and Psychiatry, Vol. i., No. 2.

[†] Teoria fisiologica della Percezione, 1881.

[‡] Influence of the Mind upon the Body, 1884.

to a minimal supply. Faintness from an overpowering emotion or impression may thus be caused.

In expectant attention, the nutrient reflex anticipates the activity, and the "dilemma period" is shortened, and more so if the thing to be seen, as a red color, is denoted beforehand. Memory recalls the red color, and the optic area is better supplied with blood, while adjacent areas are inhibited vaso-motorially.

Thus, in perception and attention, the vaso-motors are concerned largely.

Sir Henry Holland notes epistaxis increased by attention, and John Hunter was able to produce a sensation in a part by fixing his attention upon it. Unzer says: "Expectation of the action of the remedy often causes its operation beforehand." All these are connected with vascular changes. In some, this ability to recall a peripheral sensation is accompanied with increase of blood-supply and heat in the part, another vaso-motor reflex produced by cerebro-spinal centrifugal irritation. I regard this as more an unconscious motor phenomenon, the sensation and vaso-motor act being secondary. Maury, quoted by Tuke, refers stigmatization to a similar cause.

CHAPTER. XI.

MORPHOLOGY, HISTOLOGY AND EVOLUTION OF THE HUMAN BRAIN.

In several articles, published in scientific and medical journals,* I advanced and maintained what is known as the intervertebral homology. The fact of its having originated in America has delayed its general notoriety, as semi-scientific men on this side of the Atlantic prefer, as a rule, to credit startling things to our transatlantic pundits, or at least to postpone notice of American ideas until forced to acknowledge their value. It is the same spirit which, among the less intelligent, affords the " dudesque " Anglomania.

The simplest spinal cord is owned by the amphioxus, a form

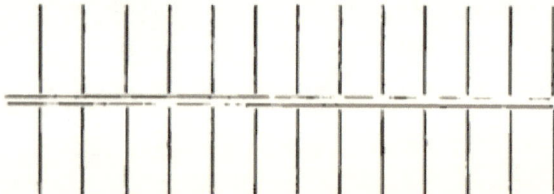

lower than the lamprey eel. This fish-like animal has no brain. Extended the length of the body is the cord, and nerves enter it dorsally and ventrally; the second pair of nerves at the head end, (Fig. 23) pass caudally, according to Owen, but Huxley does not describe them as so doing. They merely

---

* *Journal of Nervous and Mental Disease*, October, 1879. "Cerebral Topography;" April, 1880, "The Sulcus of Rolando and Intelligence ;" October, 1880, " Plan of the Cerebro-Spinal Nervous System." *American Naturalist*, January and February, 1881, " Comparative Neurology;" July, 1881, "Origin and Descent of the Human Brain." *Chicago Medical Journal and Examiner*, November, 1880, "Cerebral Anatomy Simplified."

appear to be slightly more complex in their anterior distributions. Those uppermost in the diagram are, along the back, sensory nerves, the lowermost being motor.

The cord of the lamprey (*Petromyzon fluviatilis*) is quite rudimentary, but a distinct brain presents itself in this case for analysis. We find certain intumescences at the head end which can be represented schematically thus:

The real appearance of these ganglionic swellings, for such they are, resembles the embryonic fusion of cerebral and spinal ganglia. Notice that in this low vertebrate form these enlargements of the sensory or ingoing nerves occur at the head.

A teliost fish, the *Trigla Adriatica*, affords an example of these same enlargements appearing all along the spinal column:

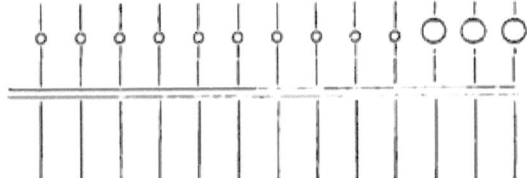

The *lateral* fusion also between these ganglia in the head end, occurs among the intervertebral in *Orthagoriscus mola*.

Taking a general survey of the piscine and amphibian brains, we find, in many, these ganglia well defined as rounded, symmetrically placed bodies (lepidosteus, amblyopsis, leuciscus), while in others, these lobes are distorted by elongation or cramping in all directions (sturgeon, sharks, chimæra), and in still others, some of the lobes are pushed below the usual site (cod, herring, perch). Of necessity, the ventricles must often be partially or wholly obliterated, showing the inexpediency of making use of ventricular passages in homologizing.

This crowding together, fusion and distortion of ganglionic lobes, obtains throughout animal life, and the olfactory lobe is often so closely fused with the prosencephalon as to afford us no line of separation. The corpora bigemina, which lie upon the upper surface of the brain in reptiles, are succeeded in birds by these bodies being thrown down to the sides and base of the brain, crowded there by the greater relative size of the superior lobes.

The intervertebral ganglia which develop on the afferent nerves of the higher vertebrates undergo great development within the cranium, and by lateral crowding together the median line of separation is obliterated, giving us the large central lobe of the shark and birds. Two or more of these ganglia, as was noted by Davida, may develop upon the same sensory strand. The subsequent lateral lobes of the cerebellum can be resolved either into secondary or primary ganglia, or a mixture of both, certainly the vagus tubercle of the fox shark is in all essentials the pneumogastric lobule of man's cerebellum, the flocculus.

Thus it appears that by the pressure together of a number of these posterior root swellings a cerebellum has been formed. The cerebellum is now generally conceded to be a coördinator of sensation for cranial sensory nerves, and how can it be otherwise from this view? By this coalescence of intervertebral bodies it follows that sensations passing in from a variety of points must be distributed to a wider area of central points in the medulla and spinal cord. This explains why injury to the lateral lobes may occur without manifestation of the lesion, and why a disorder of the central part, or vermis, produces an altered gait. The main bundles of ingoing nerves are gathered in the latter region, while the plexus of fibers in the latteral lobes afford many avenues for impulse passage, other than those injured or destroyed. The original globular appearance of the lobes composing the cerebellum may be well made out in most quadrupedal forms, but as we pass to man we see that these lobes have become compressed into laminæ.

All tubercles of the vertebrate brain fall within the category of the intervertebral, a notable instance being the Gasserian ganglion. Mr. A. Milnes Marshall,* in an article "On the development of the nerves of the chick," shows plainly that the olfactory nerve must be considered homologous with spinal nerves, for it is similarly developed and in no way differs from a spinal nerve. Nor does the comparison rest here, for the lobe (not bulb) of the mammalian olfactory may be seen to be developed between the central tubular gray and the periphery, just as is an intervertebral ganglion. As to internal structure, the law of differentiation shows that subsequently acquired differences are no arguments against original derivation, for what can be more unlike than bone and cartilage, hand and foot, skull and vertebræ, and yet one is a developed or differentiated form of the other.

Thus the mammillary eminences, the optic and post-optic lobes, were originally intervertebral, and the olivary body embedded in the spinal gray is another, related particularly to innervation of the tongue. It is large in the parrot, and has relation to the ability of that bird to articulate. But the most general interest centers in the large mass of nerve fibers and cells called the cerebrum. In the ornithorynchus it is smooth and simple in form. The beaver has an unconvoluted brain, which also shows at once the folly of attaching psychological importance to the number and intri-

* Monthly Microscopical Journal (London), October, 1877.

cacy of folds in animal brains. With phrenology, which locates bibativeness in the mastoid process of the temporal bone, and amativeness in the occipital ridge, the convolutional controversy must die out, as did the old so-called science of palmistry which read one's fate and fortune in the skin folds of the hand.

The most noticeable change in form, as we pass up the scale of mammalian life, occurs in the production of the fissure of Sylvius. In most quadrupeds the olfactory lobe fills up largely the anterior part of the cranium. As the smelling sense diminishes, this lobe degenerates to a mere tract, and the frontal lobe of the brain increases in size, lifting the forehead into a vertical plane. The medulla oblongata is pushed forward to a less oblique angle with the front of the brain from lemuridæ to chimpanzee and man; the frontal lobe pressure covers the cerebellum with the backward progress of the occipital lobe till finally the occipital forms the temporal by curling under and forwards, forming the Sylvian fissure. These stages of progress are evident in the horse, elephant and human embryo. Often, in idiots, we find through want of development of this frontal lobe, that ossification takes place in a plane inclined at an angle corresponding with that of lower animals, and the cerebellum is uncovered. This is an adaptation of the skull to its contents, which, however, does not always take place.

There are other elements at work to cause the skull to de-develop normally, or even enlarge it abnormally, as, for example, an accumulation of water in the ventricles will change the relative positions of the cranial bones to such an extent as to give to the hydrocephalic idiot "the front of Jove."

While the ontogenetic stages of development resemble strikingly the forms mentioned by Hæckel, the nervous system is not apparent in the embyo until we reach the ninth or acranial stage, after this the cerebral vesicles rapidly develop and resemble in general the cyclostome stage. Just as the sharks and mud fishes possess the intervertebral ganglion, which the hags and lampreys have not, the human fœtus, subsequent to the shaping of the cerebral vesicles, develops the posterior spinal nerve root swellings. From this point upward it is easy enough to observe that, like the condition found in the brains of marsupial adults, the cerebellum is at first uncovered, then by frontal lobe growth the temporal lobe is formed, as in man and apes.

The human brain, like everything else in the universe, has been evolved from a simpler condition, and by a study of the forms through which the cerebrum has passed to its present complexity in man, we have been enabled not only to see the reasons for changes in appearances, but to simplify our methods of instructing students. I believe, with Prof. Burt G. Wilder, of Cornell University, that no medical student should be allowed to dissect the human cadaver until he has previously familiarized himself with the anatomy of the cat. If previous to this, even, he should study the crayfish, with Huxley's recent popular work on that crustacean as a guide, he will find that he not only has removed vast difficulties from his course of study, but will be able to continue through life an acquaintanceship with anatomical and physiological essentials, of which our present methods of teaching enable the average student to retain only a smattering, however industrious and earnest he may have been. In short, we cannot master human anatomy without a knowledge of comparative anatomy.

The primitive typical form of the mammalian cerebrum is as simple as this figure, an oblate spheroid:

In some fishes and birds it is globular, and therefore a still simpler form.

Rotate an ellipse about its shorter axis, and then cut the figure so produced into halves in the direction of its long axis, and the low form of the hemisphere from which we start is obtained. We have now a flattened surface which has been created by the two right and left lobes pressing against each other as the brain grew faster than the skull. In some fishes the original spherical appearance of the two lobes may be seen. The fissure which separates the two hemispheres is called the *Great Longitudinal Fissure,* and this cleft is along the inner flattened faces of the hemispheres, each of which surfaces is known as the *Median Surface,* The *External Surface* is the outermost and uppermost part. The *Basilar Surface* is simply the lowest or under part of the cerebrum.

Draw two parallel lines lengthwise upon the external surface. These lines are sulci, or little fissures produced by the folding in of the soft brain tissue, as it develops more rapidly than the bones of the skull expand :

One such furrow may be seen on the hedgehog's cerebrum, the other is added in higher animals. But at the same time there appears another lobe growing in front of this; it is the frontal. The smallest frontal lobe, speaking relatively, is owned by the kangaroo, and adhering to our schematic representation is shown in the above cut. In the horse we find that these two lobes have crowded together, the frontal having grown much larger, but the original line of junction between the two is still evident in a fissure which in the elephant, monkey and man is called the *Sulcus of Rolando:*

This frontal lobe or subsequent growth also has the two longitudinal cracks in the highest mammalia. But the skull still obstinately maintains its rigidity, lifting a little on top as we pass up the scale of animals. The hemispheres press back over the cerebellum only at this stage of high development ; previously the cerebellum was uncovered by the cerebrum, and now another change begins. As the frontal lobe continues to grow, it crowds the occipital part back, and the latter cannot extend in the same direction any longer, but finds room below, in the posterior part of the skull, whereupon this appearance is presented :

Where the posterior lobe folds under, the temporal lobe is forming, and the three great divisions of the cerebrum are more evident—the *Frontal, Occipital and Temporal Lobes.* The first two are separated by the sulcus of Rolando, the last-named from the occipital by the large fissure, created by this folding-under process—the *Fissure of Sylvius.* The elephant's brain exhibits just this stage of development, and it is also to be seen in the human embryo.

In accomplished development the scheme of the cerebral fissures and sulci of man would be thus represented :

The temporal lobe passing forward as the olfactory lobe of quadrupeds diminishes in size and makes room for it.

But by gradual filling in of certain portions of sulci, breaks are made in their continuity and the system of folds is rendered complex. The original derivation, however, is not completely masked, for we may still trace the primitive furrows into the fully developed cerebrum of man.

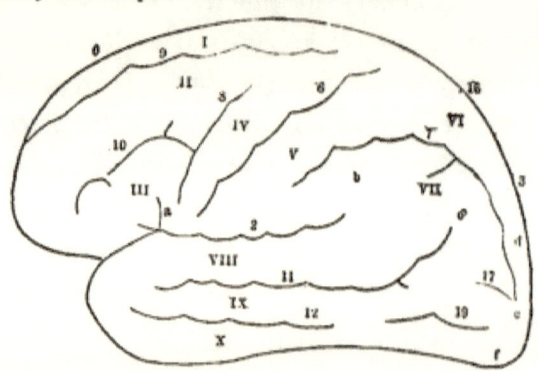

The convolutions marked I, II, III, correspond in the frontal lobe to those marked X, IX, VIII. The upper posterior convolution VI being a continuation of X, while the point c of the fissure II has been pushed back by filling in of the space anterior to it.

The deeper furrows are known as fissures, and those less deep or constant as sulci. The numbers and letters in the following list correspond with those of the figure above:

Convolutions of the External Surface.—I. Superior Frontal. II. Middle Frontal. III. Inferior Frontal. IV. Ascending Frontal. V. Ascending Parietal. VI. Superior Parietal. VII. Inferior Parietal. VIII. Superior Temporal. IX. Middle Temporal. X. Inferior Temporal.

The fissures and sulci of the external surface are: 2. Fissure of Sylvius. 6. Sulcus of Rolando. 7. Parietal Sulcus. 8. Precentral Sulcus. 9. Superior Frontal Sulcus. 10. Inferior Frontal Sulcus. 11. Superior Temporal Sulcus. 12. Middle

Temporal Sulcus. 17. Transverse Occipital Sulcus. 19. Inferior Longitudinal Occipital Sulcus.

Small parts of the convolutions conveniently designated Gyri are known as: a. the operculum (injury to which causes aphasia). b. Supra marginal gyrus. c. Angular gyrus. d. First occipital gyrus. e. Second occipital gyrus. f. Third occipital gyrus.

The median surface development proceeds in animal life as follows: With the advent of the frontal lobe, its inner face fuses with the occipital, or what becomes afterwards the parietal part of the occipital, on a line which divides the inner surface just as the sulcus of Rolando divided the external surface.

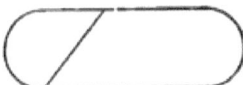

But the corpus callosum appeared in forms of life above the marsupials and prevented the backward extension of this crack. It had to pass backward over the corpus callosum thus:

And in this figure we see the corpus callosum (a broad band of fibers connecting opposite hemispheres with the *callosomarginal sulcus* just above it.

As the hippocampus major was curled under and forward by the frontal lobe pressure, the rotation of it and the fornix, together with the resistance of the skull behind, folded in the *parieto-occipital fissure*, which in monkeys extends across the external surface, but in man has been filled in upon that part:

When the calcar avis or hippocampus minor developed in
the ape's brain, another fissure, the *calcarine*, appeared:

The fully developed Median surface of the cerebrum is represented in this cut:

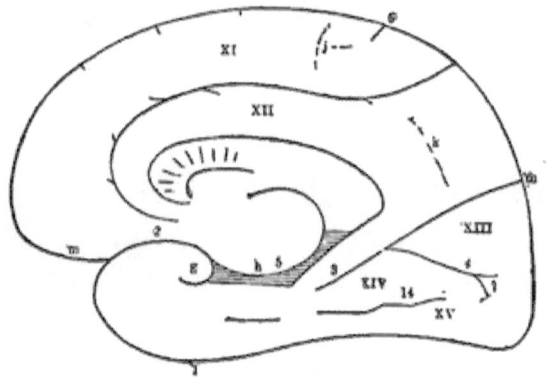

Fissures and Sulci.—6. Termination of Sulcus of Rolando.
16. Callosso-marginal Sulcus.   3. Occipito-parietal Sulcus.
4. Calcarine Fissure.   5. Hippocampal Fissure.   14. Collateral
Sulcus.

Convolutions and Gyri.—XI. Marginal Convolution. XII.
Fornix Convolution. XIII. Cuneus Convolution. XIV. Me-
dian-Occipito-Temporal Convolution.   g. Uncinate (or hook)
gyrus.   h. Dentate gyrus.   j. Paracentral gyrus (lesion here
always causes paralysis).   k. Præcuneus or Quadrate lobule.
l. Descending gyrus.

The Cuneus Convolution (XIII), has recently been discovered to be an "absolute" center for sight impressions.
Injury thereto causing optic sense derangements.

This last cut shows the convolutions and furrows on the Basilar Surface :

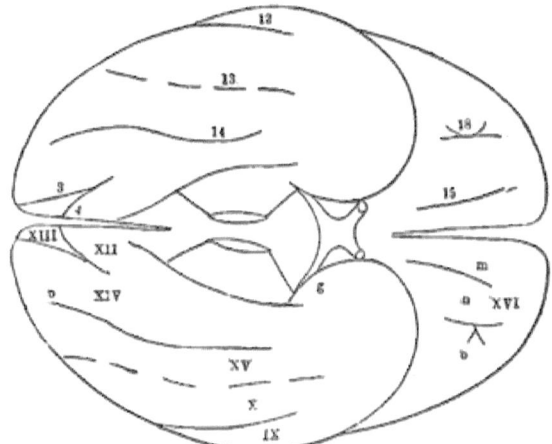

Some of these parts have been mentioned under the heads of other surfaces, as they may appear upon two surfaces. Those most clearly basilar are :

Fissures and Sulci.—13. Inferior Temporal Sulcus (which is an extra fold in the human brain, due to the same causes that created the original external furrows—want of cranial osseous expansion). 15. Olfactory Sulcus (produced by the olfactory lobe, which in man has dwindled to a mere tract). 18. Orbital Sulcus (which lies directly over the orbit and is created by lateral and frontal pressure).

Convolutoins and Gyri.—XV. Lateral Occipito-Temporal Convolution. XVI. Orbital Convolution. m. Gyrus Rectus (inwards from and along olfactory sulcus). n. Middle Orbital Gyrus. o. Lateral Orbital Gyrus.

In addition to such fissures and sulci as are of constant appearance, a great number of sulculi or lesser furrows of an inconstant nature are also present ; these of course cannot be enumerated. In young children and imbeciles the sulcus of Rolando will be found much farther toward the front than in fully developed brains. This is because the frontal lobe has suffered arrest of growth. This sulcus appears very close to the front in idiots, and their retreating foreheads are cranial adaptions to this defect.

The position of the cerebellum and its recognizable phylogenetic changes may be easily traced through the vertebrata generally, but the lobes superior to it undergo a variety of distortions and changes of position, for the solution of which we must resort to schematic views.

Given, a series of tubercles which shall from behind forward represent the lobes of the brain, as follows:

1. Posterior pair of tubercula quadrigemina.
2. Anterior pair of tubercula quadrigemina.
3. Epiphisis cerebri.
4. Mammillary eminence.
5. Olfactory lobe.
6. Cerebrum.

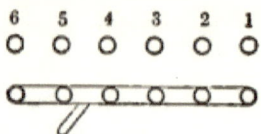

The gray secondary of each being united by commissures, the afferent and efferent. The first of these commissures it will be most convenient to follow through the developmental gyrations as apparently connecting the under surface of each lobe, but in reality connecting the secondary segments pertaining to each, as optic thalamus, tuber cinereum, olfactory ganglion and corpus striatum.

1 is connected to 2 and 3 by the upper end of the brachium conjunctivum, 3 to 4 by prolonged habenulæ, 4 to 5 by fornix, 5 to 6 by hippocampal fibers, tractus Lancisi and gyrus fornicatus (the latter principally). In the case of a fish with optic lobe (2) developed covering the other tubercles, the course of

the commissures and relative mass appearance would be thus:

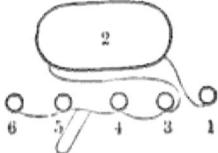

Bird, as pigeon, with cerebrum developed covering 1 to 5, the optic lobe being pressed to one side.

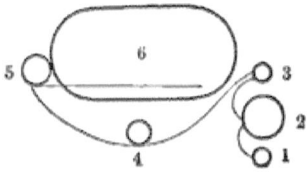

The following appears to be arrangement of the brain of the fox shark, with lobes equally developed. I think the main mass must be the optic thalamus, with the quadrigeminal bodies fused on its surface (this latter feature not represented here).

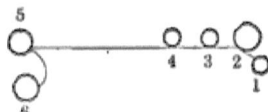

Ths form appears in mammal with large olfactory lobe and cerebrum.

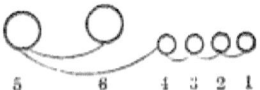

This condition is presented by an unconvoluted, brain such as the beaver's, which is but faintly fissured.

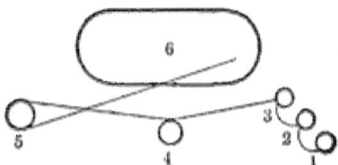

An illustration of the gradual appearance of the Sylvian fissure with the hippocampal formation, is attempted below:

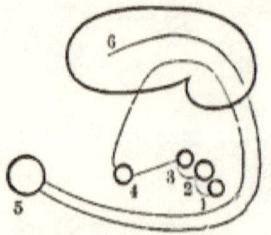

The last cut represents the Sylvian fissure formed as in man, with the accompanying fasicular distortions:

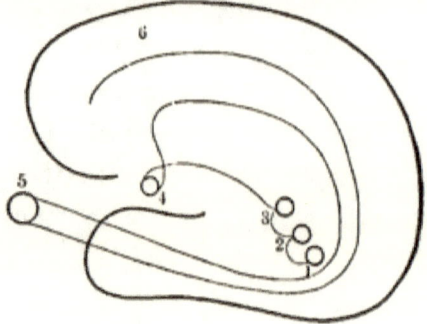

The growth of the frontal lobe in proportion to the intelligence of the primate individual augments this creation of temporal. Many of the longitudinal sulci of the quadrumana fold over and under with this advancement of the occipital into temporal, and the parieto-occipital fissure on the median face of the cerebral hemisphere is doubtless created directly by this bend, and the calcarine may also owe its origin to this change. A variety of causes combine, however, in fissure formation, aside from those mentioned.

Kölliker, Key, Retzius, Schwalbe, and other histologists concur in the intimate structure of the intervertebral (sometimes called cerebro-spinal) ganglia exhibiting medullated sensory nerves passing through without anatomical connection, and that some non-medullated fibrils occur. Fibrils pass through the ganglia and end in them. The Ranvier "T" cells seem to me to be comparable to the Purkinje cells of the cerebellum, and as every interposition of sensory masses or cells such as these must delay or destroy impressions, why can we not consider the intervertebral ganglia and Ranvier and Purkinje cells as organs for inhibition? We know that such organs exist and that it is just as important that blunting of impressions should occur as that others should be keenly felt.

It cannot be denied that a fibril connecting with a ganglion cell isolated from other connections can have no associated relationship with the cell community which go to make up the organism.

Let us assume the monopolar nerve cells as serving shunt or inhibitory purposes; and the consistency of the supposition will appear further along.

Let it be understood that it is not necessary to regard the cerebrum, cerebellar laminæ and tubercles of the encephalon, aside from the Gasserian ganglia, as *precisely* intervertebral, derived from a primitive spinal set of segments. Huxley disposes of the cranial bones derivation controversy by showing that the skull and the vertebræ arose at the same time, and hence while there might have been a cartilaginous tendency toward segmentation, there was no necessity for the previous vertebræ formation of the head bones in full. Precisely thus the intervertebral tendency was converted into the formation of cerebral lobes, and modifications proceeded rapidly in them.

Gray matter forms exteriorly, because the interior is filled with fibril plexuses which cohere in fascicular relations with strands passing through. Gradually the cortex is encroached upon with fibrils and inhibition cells.

The arcuate fibrils pass into relation with the corona radiata as fast as reason passes to instinct, doubt to determination, hesitancy to reflex action, through training, learning, etc.

Let the figure below represent independently acting olfactory and optic ganglia, with fibrils passing through and ending in them, and containing inhibition cells:

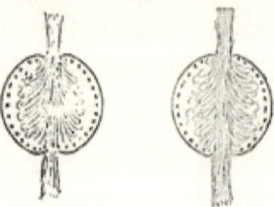

Next, by contiguity and the law of simultaneous associated action, as when a sight and smell occurred at once, uniting fasciculi would be built up from the plexiform and the association of the compound impression would occur thus:

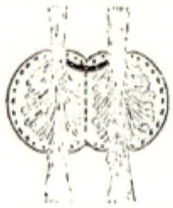

Thus, arcuate fibrils unite in the various regions different senses, or nerve inhibitory cells finding an outlet are connected with distant areas.

The occipital lobe contains the cuneus gyrus in which the optic filaments abut.

As this association occurs step by step, changes proceed. A ganglion, originally intervertebral, such as one of the anterior tubercula quadrigemina, may, by association, hand over its functions to the cuneus region of the cerebrum. Cells in this forming optic associations with others more anteriorly or elsewhere, account for the heretofore perplexing mixture of absolute and relative brain function areas.

I do not think I can make this clear to those unfamiliar with what has been done lately in cerebral physiological and patho-logical research. In a rough way, it may be stated that mainly through Exner's investigations, what he calls absolute centers for the symbolic field cluster in the region of the sulcus of Rolando, an auditory center is at the end of the Sylvian fissure, an optic in the cuneus, etc. But relative centers or scattered places in the brain occur, where injuries *sometimes* induce troubles referable to other faculties than those of the absolute centers upon which these relative centers rest.

Follow an optic thread from the optic nerve to the optic lobe, thence to the cuneus, thence by an association with a cell further forward. While the main bundles may end in the cuneus, the encroachment by association upon another center may be such as to make a relative center in another part of the brain, according to the habits, learning, occupation, ideas, etc., of the owner of the brain.

By simultaneous excitation of two sense centers in these lobes, sight and audition, or audition and deglutitory tactile, cohered. Thence over radiating fibrils to the medulla or cord for motor reflex effect.

All cerebral nerves may be regarded as sensory. As soon as an *etape* cell was overcome it became "motor" and en-larged. From this association of absolute centers sprung up a dependency of the areas, one upon the other, so that where formerly the districts were separate, by development fusion, etc., distinctions were broken down and the direct conveyance to a relative center resulted, though the absolute area largely remains, due to the recurrence of olden impressions. To fa-cilitate transfer, the tactile, auditory, optic, etc., centers are dif-fused and lie in adjoining layers, with probably no particular function to each layer, except such as would occur through adaptability of inhibition cell sizes.

The hippocampus major could have been built up through deglutitory and olfactory senses cohering. It is well known how important the smelling sense is in many of the lower animals.

In the simiadæ, as the optic sense arose in importance and largely took the place in intelligence of the olfactory, the cuneus fibrils could have and probably did in their definition give weight to the hippocampus minor for deglutitory association. Such large strands might appropriately be considered as having as extremely important functions as these assigned them, and yet their ablation need not point to their functions.

The ages during which the major fibrils have been in use, and the comparatively recent formation of the minor, account for the relative sizes, and the obsolescing features of the major may be advantageously compared with the good definition of the minor.

Goltz* concludes, from experiments on dogs:

1.   That the cerebral cortex is the organ of the mind; that the removal of large pieces of each half of the cortex diminishes intelligence.

2.   That it is not possible through destruction of a section of the cortex to cause paralysis of the muscle.

3.   That it is also impossible by a destructive lesion of the cortex to cause a permanent loss of activity of the senses. The animal has all his senses. After the removal of large pieces of the cortex there ensues weakness of the perceptive faculties.

When a tract conveying an impulse in brain or cord is injured, there is an attempt at vicariation by the contiguous gray matter; where this is also injured, as in multiple cerebrospinal sclerosis, the choreic motions may occur, and in ataxia the gray is long in conveying impressions. Mental action may be similarly involved and in the same diseases.

---

* Foster's Journal of Physiology, vol. iv. Nos. 4 and 5.

Diffusion precedes definition, and the latter in some parts is established at the birth of the child, but alternately higher and higher diffusion and defined tracts arise as the infant grows older. Until the cerebral diffusion apparatus arises in the human being it cannot think, as we understand thought. Its lower reflexes are defined but not coördinated. Memory begins in the intellectual sphere of the child about the third or fourth year, but it is so gradually evolved from birth with other lower memories as not to be determinable when first established.

It would require a separate and larger volume to fairly enter into the details of the innervations and changes wrought in the brain in the evolutionary scale. We can generalize here sufficiently by recalling Spitzka's dictum, that as the association fibrils multiply, so does intelligence advance.

The diffusion apparatus constantly tends to definition and the higher sense connections to usurp the place of the lower. Were it possible for such a thing to be accomplished in full, psychic life would cease, for the intellect depends upon its advance. Change *must* occur, and no sooner do higher diffusion associations result in defined automatic reflexes, than a superior condition of things, or retrogradation, or at least a different state of things must ensue. The higher the animal or man, intellectually, the more the cortex gray is encroached upon by fibrils; but were this process to be accomplished to full conversion into fibrils, the animal or man would become a mere intellectually lifeless machine. While life tends toward complete automatism, the safety of the mind lies in the vicissitudes which prevent it.

Another fact which bears upon my nerve-cell theory is, that the processes of nerve cells increase in number in an ascending scale of life. As the inhibited sensations for which the organism finds no use cohere with other impressions and cause reflection, wonder, or a search for an outlet for the diverted impressions ensue. The nerve cell processes would multiply, and impressions eventuate in movements which the animal, either mistakenly or correctly, considers serves some useful purpose.

Volitional impulses, according to Vulpian,* are confined to the lateral columns in the cervical regions, though in dorsal and lumbar regions in anterior columns as well.

Schiff distinguishes between tactile and general sensibility, and locates them in gray and white parts of the cord. Physiologies, such as Foster's, detail these matters satisfactorily, and reference should be made to such works for what cannot be embraced here.

As the fibrils are best furnished in the region of the cord, and the plexus fibrils being drawn upon to form the segmental nerves first, the longitudinal fibrils will cling to and form from the gray cord and follow its length. The shock of a nerve impulse is not thus sent direct from the brain to the muscular segments, but to the levels of the cord and thence to the muscles over their motor nerves. Hence, all motor influences originate in the gray, otherwise tetanic spasms, epilepsies, would occur as a rule instead of exceptionally. The sensory longitudinal columns arose by the same laws of definition. With the more intense and frequent impulses, in head to tail directions, the plexus fibrils arrange themselves longitudinally in lines of least resistance.

By adjustment it seems that a nerve arranges its molecules so as to require a stimulus above a certain rate, and below another to react. Different nerves are placed within reach of certain stimuli mainly by conjoined apparatus and placed be-

---

* Syst. Nerv., lec. xvii.

yond the reach of other stimuli, so that general impressions are disregarded in favor of special impressions. The internal central connections then only operate when the stimulus to which they are adjusted is reached.

The development of the symbolic field in the brain is what raised the forehead angle, and also enabled man to cope better with his enemies and survive.

CHAPTER XII.

PRIMARY ACTIVITIES.

Sensation. Approximate definitions only are possible in a world where everything is relative. Metaphysicians waste much time in seeking the absolute in everything. Physiologists and the fairly educated are satisfied with approximations, and the rest of mankind must be content with the knowledge that a perfect definition never existed.

Sensations are conditions of the molecules realized in consciousness, and as consciousness is necessary to sensation, the former is evoked by the latter, and physiologically may be considered identical with it. Charles Darwin confessed his inability to understand what writers generally meant by consciousness. Many, like Brown-Sequard, were perpetually entangling their readers with the two terms used synonymously and separately. Schiff, and others prefer the words general sensibility in place of consciousness. This serves generically to include special sensation.

Condillac properly derived all mental processes from sensation. James Mill reduces the phenomena of mind to sensation, ideas, and the association of ideas: "In the physical world there is only one fact, sensation, only one law, association," he wrote.

Consciousness is also used in the sense of cognition.

The relative conditions of the organism begin in the relative activities of the cell molecule, wherein motion and sensation are identical. The end of sensation or conscious feeling being to produce molar motions, which render further consciousness possible and conserve the life of the organism, step by step the activities develop from indefinite to definite, from indeterminate illy coördinated motions to determinate coördinate motions. From those illy fitted to cope with the environment to those which through more sensual reflexes to higher emotional expression and finally intellectually directed motions enable the animal to admirably adjust itself to the environment.

Sensation and memory are at the base of all mental activity. The molecular organic motions and through adjustability the repetition of those motions forming the lowest stage of conscious feeling, the beginning of which we can never remember, but which has evolved from the molecular activities and relapses into unconsciousness as fast as perfect adjustment to the environment is made.

The law of relativity and the fact that consciousness is proportioned to change, and fades with invariability of impression, are thus connected. The constant readjustment is the condition of consciousness and life.

Consciousness gradually evolves into what we call mind, which has sensation and memory at its foundation, and mind is divisible into Feelings and Cognitions with their revivabilities or memories.

The primitive molecular dissonances or painful states constituting feeling, desire, or sensation, were Hunger and Locomotory desire, the latter being consequent upon the first. As consequences of both of these came the desire to Excrete and Sexual desire, both being in the amœba identical and intimately associated with the Hunger desire. These primitive desires were painful in proportion to their being ungratified, and pleasurable relatively if previous gratifications were denied them. All of these desires would cease to be conscious states when privation and satisfaction were so balanced as to make supply and demand equal. In all stages of life we find this true. An easily obtained living ceases to afford pleasure. Degradation ensues upon absence of necessity for exertion. Parasitism in animals, as in men, extinguishes higher differentiated abilities, and as with the "nobility" of monarchical governments, degeneracy is inevitable upon constant receipt of *quid* without rendering the *quo.*

Upon these primitive desires, sensations or feelings (it is of no consequence what they may be called) may be erected all the subsequently acquired molecular activities evinced in molar activities.

Taking consciousness as in one of its primitive stages (we cannot say its lowest, for we know nothing as to beginnings) represented by the atomic and grosser motions of the amœba, co-extensive with sensation such as the amœba may possess, step by step with the correlations of the cells, we reach the highest consciousness known to us, that of man. The consciousness of the lowest man is like that of low animals in general, and that of highly differentiated animals, such as the chimpanzee, approximates that of the average man who, through circumstances and want of training, bothers his brains about little else than immediate creature comfort. From what has been considered mere automatism to the consciousness of impressions, such as the prick of a pin, thence to "consciousness of the ego," Schiff * has demonstrated degrees marked by heat evolution, and Prof. Herzen † shows that central acts accompanied most vividly by consciousness are those which require a more extended decomposition and cause greater calorification, and that consequently the intensity of consciousness is in direct ratio

---

* Archives de Physiologie, 1869, Nos. 1 and 2.
† Journal of Mental Science, London, April, 1884.

to the intensity of the functional disintegration, and inverse ratio to the facility and rapidity of central transmission.

The metaphysicians' quandary over the ego consciousness comes from their not duly considering the fact that each of us is only conscious of himself objectively. We are conscious even of our clothing, as is very evident to others, according to its character. You know yourself objectively by the impressions your objective self make upon your subjective consciousness, and as Maudsley says, such considerations have nothing to do with man in general, for no one but the educated white man, as a rule, would cogitate over such a matter. Man feels, he knows not why, and seldom does he care or seek to know why. All his impressions have an objective origin. Even his thought is an effect of a chain of antecedent extrinsic causes.

We cannot deny consciousness to every atom of the universe. We have reason to assign it to organic matter, especially living protoplasm, as that composition of molecules which culminates in the gray matter of the nerve centers.

Hence, so far as animals are concerned, consciousness in different degrees resides in every cell of their bodies, but more especially in the neurogliar nerve centers.

Memory is simply a repetition of a sensation. Through recurrences of impressions the molecular adjustment of the organism results. Each recurrence is consciously or unconsciously remembered as similar to the original impression. Associated molecular disturbances, no matter how induced, whether by conditions of the blood or vibrations approximating those of the impression, suffice to recall or reproduce it. For example, a sensation may be likened to a note produced on a violin-string by the bow being drawn across it. If the note be reproduced by any other means, such as vibration in harmony with another instrument, such reproduction may be likened to memory. In the insane, these revived sensations, through absence of ability to make mental correction, are often mistaken for the real ones, and are the bases of hallucinations and delusions.

When one forgets, for the time being, something he wished to have remembered, the excitement of other recollections or brain activities interfere with the process. The vaso motor supply in certain regions relatively obliterates the activity of other regions. Memory flags when the circulation is bad or low, showing how the blood is important in the revivification of adjustment appreciation. The instability of the molecules of the amœba would confine its memory possibilities to food relations.

When portions of the brain are destroyed, other mental activities *may* adopt new channels, but memory, depending upon the integrity of the part usually impressed, cannot, when destroyed, as a rule, be reinstated in such ways.

Excitement consists of every degree of activity, from the molecular movements, which constitute sensation of all grades and intensities through all bodily activities, to the violence of maniacal furor. It is manifested in all sensory and motor events, but is customarily classed among emotional states, though it would be an error not to recognize it in other activities, mental especially.

Excitement may be pleasurable, painful, or unconscious. It may also be definite, or indefinite, depending upon circumstances, and whether established or diffuse outlets for its expression are found.

Pain and Pleasure are fundamental conditions of dissonance and consonance, and cannot be confined to emotional states alone. The law of relativity is operative in creating degrees of pain and pleasure, and often changing one into the other.

Sleep and its Phenomena.—It is well demonstrated that in sleep the brain is comparatively anæmic, and there is during sleep a necessity for maintaining a minimum of blood there. The circulation is slowed and equalized throughout the body; an evenness of reparative processes proceeds. During activity the different parts of the organism have been unequally excited and an excess of waste over the instant repair has accumulated to make further activity disagreeable, if not painful or impossible.

In the amœba the quiescent plethoric state facilitates full assimilation. Imagine one cell of the morula form badly nourished through over exercise, and another cell congested, the remainder in variable states of nutrition and activity. The equilibrium of the cells can be restored by ceasing, as far as possible, activities of the organism which tend to hypernutrition of one cell and diminished supply to another. The pabulum diffuses itself. Where one cell, through great hunger, kept up an active demand upon the others, sleep or quiescence of all would be disturbed until the activity diminished. It is for this reason that sleep does not always appear when needed. The laborer whose brain has not been particularly hyperæmic but whose muscles are tired—exhausted through their cell waste, with withdrawal of most sense stimuli, as light, sound, etc., " drops off quickly to sleep." Not so the student whose brain is engorged with blood. In spite of stimuli withdrawal, wakefulness persists. Active or passive hyperæmia takes a longer time for bodily equilibration. Frictions, baths, etc., assist this by diffusing the blood supply, and probably a nocturnal meal would facilitate release from congestion of the head by "draining " blood to the stomach; but excessive digestive function going on is not rest, and it is only by practice that this is tolerated or adjusted to.

A not very hearty meal at night might assist the student.

The growing child needs more sleep than the adult, for the reason that the growth processes demand long intervals of freedom from activities directed in other than growth channels.

The aged need less sleep, for involution renders attempts at reparation largely abortive. The cells are growing less capable of activity, they need less repair and could not assimilate so freely as when young, and there is a quantitative lessening of the cell number also. The faculties grow more obtuse and are not using up material, hence what little reparative need there is, a few hours furnishes.

Animals differ as to their needs. Some sleep by " snatches," others hibernate, when, during a season of dearth, food is not easily obtained; as sleep minimizes the tissue loss it is resorted to. Other animals and men, through stress of circumstances, as when living in polar regions, grow to disregard the presence of light during sleep, by a process similar to that which enables one to sleep with a lamp alight in the room. Nocturnal animals also, who have adapted themselves to prowling, acquire the ability. Napoleon is said to have possessed the power of dropping off to sleep in spite of activity in the neighborhood.

Unpleasant excess of heat stimulates to awakening, and one so over stimulated will fall off to sleep immediately on being cooled off.

Huebel* describes forced hypnotism in cold blood vertebrates by holding a frog, without paining it, excluding light, sounds or any other nerve stimuli; the animal sleeps for hours even in a constrained position. This indicates what can be deduced from other matters, that sleep is a withdrawal of vibratory sense stimulation, to which the higher centers re-

*Pfluger's Archiv.

spond, and as the spinal cord sleeps by resting the muscles, so do the optic, auditory, etc. centers by resting peripheral organs.

The heart muscle is compelled to rest in the interim of beats, and it is calculated that it receives eight of the twenty-four hours for this purpose. Then with the lessened heart action during sleep, repair to the organ is further favored.

The stimuli withdrawal from cerebro-spinal nerves instantly (if other things are equal) stays the nutrient reflexes. No more blood in excess is sent to the head, the fall of blood pressure in the medulla drops the heart beats to a lesser number through pneumogastric action, and the equalizing pressure begins. Where an impression, emotion or other brain activity is too great to allow its blood to depart from the region of the brain involved, then diverting the mind by calling to it a different train of thought, through a chain of antecedent causes, may, and sometimes, not always, does act derivatively to distribute the blood and permit sleep.

Normal sleep should be dreamless, but when dreams occur it is through some cell activity about the body which has been aroused and suggested feebly or strongly through the particular area of the brain to which the blood is sent or drawn. The fibrils and sensory cells, thus thrown into erratic activity, revive the vibrations to which they were subjected in the past, and as these blood interferences with sleep approach more the methods normally obtaining during waking hours, so will the vividness and consecutiveness of the dream or nightmare from congestion be. Where first one part and then another part of the brain is thus supplied with blood, through internal or external stimulation, the most absurd association systems are temporarily formed, and with pathological conditions to favor this, through extinction of the ability to correct such impressions through other senses, or by reason in waking states, in-

14

sanity, with delusions, hallucinations and illusions, is induced.

False association is at the root of many forms of insanity, if not all forms.

"Two of Vierordt's pupils, Mönninghoff and Piesbergen, have made the depth of sleep the subject of investigation. The depth of sleep is proportional to the strength of the sensory stimulus necessary to awaken the sleeper, that is, to call forth some decisive sign of awakened consciousness. As a sensory stimulus they made use of the auditory sensation produced by dropping a lead ball from a given height. The strength of the stimulus was reckoned, in accordance with some recent investigations of Vierordt, as increasing, not directly as the height, but as the 0.59 power of the height. For a perfectly healthy man, the curve which they give shows that for the first hour the slumber is very light; after 1 hour and 15 minutes, the depth of sleep increases rapidly, and reaches its maximum point at 1 hour and 45 minutes; the curve then falls quickly to about 2 hours 15 minutes, and afterwards more gradually. At about 4 hours 30 minutes, there is a second small rise which reaches its maximum at 5 hours 30 minutes, after which the curve again gradually approaches the base line until the time of awakening. Experiments made upon persons not perfectly healthy, or after having made some exertion, gave curves of a different form."[*]

The withdrawal of stimuli voluntarily, through closing the eyes, is only partial; noises are heard, and the blood does not immediately fall to a minimum in the brain; hence, in most cases, sleep cannot appear until time has passed. As the nutrient, sympathetic, or vaso motor reflexes are less and less called upon, an ebb of blood from the brain finally admits of obtuseness to noises, etc., when they do occur. With the restoration of general cell-nutrition the desire for activity increases, and the reflexes are easily provoked. The nutrient

---

[*] Zeitsch. f. biol., xix. 114.

reflexes of the brain now begin to send blood there upon stimulation, and the man awakes with the increased noises and light of day, or upon slighter provocation, if these are absent.

The condition of the health and many other matters vary the phenomena, but, as a rule, the foregoing is about the process.

CHAPTER XIII.

DERIVED ACTIVITIES, MAINLY THE EMOTIONS AND THEIR EXPRESSION.

Pain or pleasure to an amœba could only be judged of objectively as excitement, increased activity, and the fundamentality of excitement is shown in all animals while experiencing either or both feelings. All emotional activity is excitement, even if not expressed in any way. The suppression of the emotion itself is an excitement.

Wundt[*] affords numerous proofs of the highly developed animals standing nearer to man than to lower animals of their own genus even, hence it is not proper to regard the emotions, etc., of the dogs and apes in any other way than *with* our own, not as leading up to ours. Many of the higher mental acts of such animals have been developed at the same time with our own, and for the same reasons. We can simplify our treatment of the subject in this way, and work closer to first principles.

The amœba is unable to express the difference between that which is dissonant among its molecules and that which is consonant with their workings.

Pain or pleasure activity in synamœba or planeæda could not be judged apart. In the latter, ciliary motions would be induced by either, but with the worm, provided with a low nervous system badly coördinated, efforts to move would be made by both feelings. The pleasurable feeling would be referred

---

[*] Vorlesungen über die Menschen und Thierseele. Leipzig, 1863.

directly to assimilation, but in the worm, and probably lower, pain could not be referred to and induce the excretory locomotory motions. As the expulsion of CO_2 and other inert matter from the body is associated with unpleasant feelings, interferences, so all pain has reference to this, and thus in all animal life involuntary evacuations may be induced if the disagreeable sensation may not be disposed of upon determinate motor channels in other ways. This fundamental pleasure and pain excitement, ingestion and retarded evacuation, may exchange places in retarding the assimilatory and facilitating the excretory if previously opposite conditions existed.

Relativity, as Bain shows, must be perpetually kept in view in discussing the emotions, but the foundation of painful emotions is in assimilation interference, of which excretion is an associated act, and pleasurable emotions are grounded upon gratification of hunger, of which excretion is a consequence. The association in sequence is so direct in cell life that it has implanted itself below all subsequently acquired feelings as a basis.

The vulgar expressions, "good enough to eat," and mention of sounds being "sweet," show the close relationship of all pleasures to the gustatory sense. Conversely, epithets of dislike are full of references to expulsive functions. The expression of disgust is a faint semblance of vomiting. The sneer indicates that the nostrils are closed against a bad odor. In the lowest forms of life, then, all pleasure and pain had reference to the ingestive and excretory functions. Where, through too great pain, an interference with the ingestive mobility was produced, then "low spirits" would be expressed, and this mode of expression through supineness pervades all the cells of the organism when that condition occurs in higher animals. The nutritive functions are interfered with in all during mental, nervous or other cellular prostration.

Excitement in all animals, to the very lowest, whether painful or pleasurable, may be evinced in tremblings, modified secretions, perspiration in excess, whenever there are not definite means of conversion of the diffusion into evolved channels for their expression.

With the better association of the cells through a nervous system, appropriate motions can be made, the offensive substance can be expelled or suitable movements made to swallow the pleasant substance, the presence of which produced the sensations. When the muscles are arranged to facilitate these acts, associated serviceable habit, the principle discovered by Darwin,* will group pain and pleasure expressions generically into certain muscular expressions. Thus, all pain, whether bodily or mental, produces motions which were originally useful, such as contraction of the brow and closure of the eyes upon receipt of painful intelligence, as though to avoid a strong, painful light.

An expression originally serviceable may persist through habit, even when positively unserviceable, just as the pageantry and buffoonery of monarchical ceremonies persist, though the multitude, as the years advance, are inclined to look upon them with growing disfavor and contempt.

Disgust underlies all other painful experiences, and its expression indicates a desire to rid the stomach of an offensive substance. No matter what disgusts us, it has, through our muscular expression and language, this primary significance.

The intimate association of all forms of disgust with stomach and intestinal distress was fully shown by a Westerner, when looking over his hotel register, promptly depositing his just eaten dinner thereon upon seeing Oscar Wilde's name in-

---

* Expression of the Emotions in Man and Animals.

scribed in the record. An antipathy to the dudesque was coupled in this case with an irritable stomach.

Offensive odors, when strong enough to interfere with respiration through imperfect oxygenation, excite the same disgust, feelings and expressions.

Such odors as do not interfere with respiration are offensive or pleasant through association, as is well attested by animals, savages, and even civilized men acquiring a relish for what to others are malodorous edibles.

In the low forms, from amœba to the worm, we see:

Excitement, molecular and molar, as a result of the dissonant and consonant changes which constitute pain and pleasure, and faintly set forth evidences of these primitive animals experiencing

Low and High Spirits, according to whether the feelings are disagreeable, dissonant, or agreeable, consonant.

Dejection and *Joy* are merely synonyms for the above. *Despair* and *Grief* also express the same thing as dejection and low spirits, only they are more intense, and should be reserved till the senses are more developed and consciousness is more acute.

All these feelings are expressed to this low stage by excitement and not very well coördinated movements directed to achieve ends.

These diffused impressions occur in man and induce the badly coördinated motions, such as trembling, etc., for the same reason that they affect the protozoa and lower metazoa in a similar way, because, for the time being, there is no outlet for the feeling, the impression cannot, through its over intensity or its newness, find the proper manner of adapting motions suitable to it. In painful conditions this constitutes *Fear,* which can come only with the establishment of memory.

As soon as the molecular adjustment to eating is stable, the memory adjustment is aroused, and with hunger pain and memory adjustment in the protozoa came the development of what in higher animals, with nervous systems, could be recognized as fear, which, from its origination, was associated with the excretory desire or the want to get rid of troublesome interfering substances. In all animals excess of fear will act upon the excretory organs.

Fear is gradually evolved as the amœba and its descendants have frequently experienced pain. The memory of the pain (revivification of the molecular movements previously induced by a pain) imparts a vague, distressed sensation tending toward dejection, and when the feeling is revived, according to the strength of the influence in animals with nervous systems, and no adjusted arrangement for the impression in muscular contractions or means of escape, degrees of fear arise beginning with disagreeable feelings, passing through disgust and culminating in *Terror*, which diffuses the strength and often acts upon the bowels through its primal association with excretory impediment feeling.

In all cases of doubt, fear, distress, etc., there are "sinkings at pit of stomach," and in extreme cases visceral disturbance to the extent of seminal, fœces, or urine voiding.

Fear of the unknown is always the most terrible ; the inability to adjust suitable means of escape thus make anticipation greater than the reality. "Prostration of the nervous system" frequently follows this sort of dread. It is a formidable force in nature and affects all animals. It is especially poignant in savages and barbarians, and is the basis of all their superstitions They are ready to listen to any one who has a theory of the unknown, providing no theory has previously been accepted in which case the new theorist is likely to be immolated. Fear is the idol and fetich-maker, and in all countries whoever can, through established custom, successfully maintain himself as an interposing power between the frightened and the dreaded unknown, may thrive upon such claims and even come to believe that he possesses such intermediating power.

As soon as acts become adjusted to ends, and the animal knows what to do to avoid, escape from, or to destroy his enemy or the inimical disagreeable thing; as soon as easy avoidance of the hateful impression is furnished by acts of the muscles, which means that the nervous system has found its proper associations with muscles, then the *efforts* thus aroused affording the escape, in the knowledge that escape may be made,—even though the knowledge may turn out to be illusory,—the feeling of defiance, *Anger*, arises. Fear is no longer felt, but with the combative motions is felt a tenseness of the muscles, through outflow of force in definite directions over better coördinated nervous systems. (Anticipating here, somewhat, I believe such tenseness arising in muscles, as in the form of insanity known as Katatonia, *suggests* to the mind the combative feelings of this disorder). Anger supercedes fear, and as fear excitement arose because there was not an outlet for it, anger arose because the outlet in effort was possible.

Contempt may now be more safely expressed, or if *Rage* be provoked through intensification of the anger, the destructive eating propensities are exhibited in the canine tooth exposure, destruction being associated with eating, and brawlers threaten to eat each other up every day.

Hatred is the disagreeable feeling, which, having its lowest terms in disgust, has its highest in anger and contempt.

Contempt is that disposition to show disgust whenever the anger effort makes it safe to do so. It is seldom exhibited when it is considered to be unsafe. Anger effort fixes the muscles in various ways, according to the muscular structure of the animals. Some birds and other forms fluff out their feathers or swell themselves up, all draw themselves up to their full height, to appear as formidable as possible. Ears of horses are drawn back out of the way of the antagonist's bites. Fear may also cause similar movements, showing the relationship of the two. Fear will, like anger, erect the hair of some animals and man, partly from diffusion of nerve impulse, and partly through serviceable associated habit. Probably diffusion caused the first effect, and as it served a purpose in frightening an antagonist, it has, to some extent, been perpetuated. The enraged bull paws the earth; this was originally useful in the dirt being thrown up in driving off flies.

In almost all stages of excitement, diffusion opens the eyes widely, the secondary useful object being to see better,

Astonishment is thus expressed and the eyes are opened, often the mouth; the ear muscles are made tense, and the respiration is slowed, because a new impression is affecting the diffusion apparatus and a sensation is trying to find some kind of an association for its outlet, and the diffusion through the nutrient reflex draws blood from other active parts, dropping their workings to a lower grade of activity. Consciousness is given up to the wonder feeling, the trying to discriminate between the harmful or healthful properties of the new experience.

Savages and boors are mainly interested in knowing whether it is "good to eat," if not directly injurious, and the query of the "civilized" man: "How much money are you going to make out of it?" His eternal *cui bono* for *all* activity, his failure to appreciate any mental effort which will not fill his or his neighbor's intestine, shows how little evolution has done for "civilized" man in developing his diffusion apparatus aside from gustatory connections.

Anxiety is anticipation of possible harm, and is only possible in animals with a well developed nervous system with its memory registrations, for fear cannot be anticipated were there no memory of past fear or pain. The tendency to depressed exhibitions show the relationship; as there is no immediate danger to elude or fight, the anger fighting attitudes would avail nothing, diffusion of the disagreeable feeling interferes with cell-nutrition, and depression occurs, relieved by occasional starts, when, through the vividness of the anticipation, the worst seems to have been realized. Fitful, troubled sleep is of this nature, as the senses cannot at this time correct the feelings. Throwing off anxiety is the substitution of an activity which will so interfere with the impression as to make it of little effect.

The poorly coôrdinated emotions of the infant are shown in its expressing in a different way excitement for both pleasure and pain. Its pleasures are mainly, at first, *only* ingestive, its pains refer to excretory impediments and to want of food. Wonder is every moment displayed, for its cerebrum mainly consists of a diffusion apparatus with imperfectly developed higher reflex arrangements anywhere in the body. Fear next arises, then anger.

Disgust and *Satisfaction* are well shown through the signs of negation and assent, the shake of the head and the nod thereof, which Darwin has shown to be universal modes of expression in all ages and among all peoples. This was derived from the infantile nod toward the breast of its mother, and its turning its head away from it when not hungry, or when disgusted.

Disdain is a form of contempt into which *Pride* enters, the latter emanating from a consciousness of superiority based, however differentiated, upon a muscular consciousness of ability to overcome. However fallacious or mistaken such feeling may often prove to be, the belief that one is superior to another is, in such, derived from a full habit and development of muscles, or at least the supposition that such things are possessed, even though the feeling may outlive the warrant for it. In the threatening, disdainful strut of the peacock, and the sidelong, impudent glance of the bully, with his swagger, this feeling of pride is begun, and association may extend the matter to purse pride and pride of brains, the pride of, as Tom Hood calls him, "the self-elected saint." The stalk and up-lifted head and poise of the body betoken the muscular derivation of this feeling.

Ill temper is a form of anger or "touchiness" which, if un-duly exhibited, shows the animal to have been himself, or that his ancestors have been, subjected to often recurring annoyances, whereby it became, or was supposed to be, an advantage to frequently show one's teeth with slight provocation. Indulgence in this instituted it as a reflex, and we have all grades of this irritability, from constant to occasional snappishness, with variability in its manifestations. Ill health or badly nourished cell groups in the body may occasion irascibility, which is merely a faint anger effort, a threat, a snarl, and exhibits the association of inner with outer sources of irritation.

Where one is pre-occupied and a certain region of the brain is surcharged with blood, the interruption puts to work another group of brain cells, and through their comparatively poor supply of nutrition they act fitfully, irritably. Until equilibrium is established by diversion or becoming interested in a new line of thought, irascibility may continue. But this irascibility may be vented upon any and everything if something un-pleasant has happened, due to the domination of a feeling and its mastery of the organism, just as fright may control the motions, to shut out all other impressions, as when a panic occurs.

Sulkiness is offended *amour propre.*

Self-love is based upon muscular pride and its differentiations, and has a deeper derivation in the life-conserving desire to avoid pain and cultivate the pleasant. Things which gratify, be they adulation or food, whether they pander to the ideas or the intestines, when denied us awaken feelings of *resentment,* which is a mixture of hatred and disgust. Anger may be thus provoked unless fear usurps its place.

Determination is muscular effort decided upon, and even though the determination may be a mental one and have reference to an act to be performed years hence, the muscles are made rigid, often the teeth and hands are clenched.

Guiltiness is, in lower animals and man, shame or the fear of consequences, more or less direct. Detection is the thing avoided for the guilty act, and when not avoided, according to the estimation of the punishment to follow will be the depth of the shame. The punishment may be corporeal or the loss of regard of another. In either event it was desirable to escape such suffering, and in proportion as the punishment is sought to be averted by appeal to arms or through expressions of helplessness, either methods of attempted escape are useful, and even though in cases when unavailing, are resorted to through the principle of serviceable associated habit.

Surprise is the same as astonishment. The useful habit of opened eyes passing as an associated habit to all feelings of surprise, even though opening the eyes widely could be of no use. For example, when you hear a piece of surprising news you cannot understand it any better by lifting your eyebrows, but in savage and animal days the serviceable habit was formed when enemies were approaching and prey was to be obtained.

Then shading of the eyes from the sun and their dilatation were
of use. This corrugation of the brows has such a derivation
in :

Perplexity, meditation, reflection, deep thought, and in all
other states wherein nervous diffusion apparatus is concerned,
and where, as yet, no definite line of action is determined upon.
Usually with the determination, if it be complete, the " brow
clears up," the corrugations disappear and the facial muscular
fixation takes its place, characteristic of the succeeding emotion.

Attention is heightened *perception.* The organism in breathless attention is " bending every nerve " to the perception. The
reflexes are stayed elsewhere and the nutrient reflex allowed
to concentrate the oxygenating and nutrient blood upon the
points concerned in the perception. At first this may be voluntary, but as the other parts become denuded of blood, they
become less efficient as interrupters of the engaged sense.

Concentration, abstraction, even to " absent-mindedness," is
thus possible. In even the feeblest of these stages an automatic
motion, such as winding up your watch, may be made, and
the next instant you cannot recall whether you have done so
or not. Complex, as well as simple reflexes or automatic acts
are provided for through repetition, and where the tracts,
through this repetition, have been built up to require but trifling
effort to start them, the mind may be working its diffusion
tracts with conscious effort and simultaneously instinctive
motions, or those which require but little reflection to perform
may be accomplished. Darwin mentions a lady who played
on her piano while absorbed in watching her canary die, and
when it was dead she burst into tears. The unconscious piano
playing had become such through much practice, and was
comparatively instinctively performed. The tension of her
feeling prevented her from realizing that she was playing. It

is a relief to distressed feeling, as before remarked, to move, and when embarrassed, drumming with the fingers, twirling something in the hand, afford a measure of relief. The rustic scratches his head, Darwin says, " as though he felt an uncomfortable sensation there," when puzzled.

The hunger feeling differentiated in the sexual and originated that desire. Association transferred the desire to the means for its gratification. In many low animals and savages very little of the sentiment which we call *Love* exists, but many developed animals not in our phylum have, like the Wanderoo monkey, which has but one mate and dies of grief over her loss, evolved many of the feelings which the best of us attach to the sentiment. Herbert Spencer's inimitable description of the compound nature of this affection, enumerates the pleasure feelings exercised by it, such as personal beauty, admiration, respect, reverence, approbativeness in having been selected from one's rivals, the pleasure of attracting the attention of others as having been thus favored, the knowledge of possession, etc. But where Spencer speaks from observation of the truly respectable class, who, through cultivation of the higher feelings, are able to stand unconvicted of allowing the baser feelings to sway in courtship, the vast mass of the truly lower classes, with the savages, seldom couple with sexual desire any other emotion. The lowest men are very much like the lowest animals in this regard, and the refinement of sex association has developed in the highest animals as well as in the highest men. Mere conventional social position has nothing to do with this development. Brutal dispositions crop out in dukes, kings, the "gentry," etc., and savages, sometimes barbarians, frequently conjoin other and better feelings with their *amours.*

The absorbing nature of the unevolved and evolved passion

has been, next to the hunger desire, the most potent influence in the affairs of the animate world, and in all properly consti- tuted persons its domination is not to be gainsaid, for in most lives, at some time or other, if not frequently, all actions and other feelings are subordinated to it.

It is, however, found that in some men and in many women the unawakened passion may slumber through their lives, or but feebly disturb the individual.

That this is the case, can be affirmed in spite of the testi- mony of *roués*, who, through their cultivation of beastiality, come to see only their own feelings in others. In women there is a good reason for it. Through ages the dread of the conse- quences of the gratification has operated as a deterrent influ- ence. Inducements to chastity, direct and indirect, are held out to them on all sides, and as their sources of information grow broader, the knowledge of the disastrous consequences of illicit love impel them, except when peculiar circumstances set all such considerations at rest, to be guarded. This expe- diency, acting through inheritance, to repress all exhibitions of the basic feelings, eventually the repression becomes intensi- fied, abnormally so in some, to the extent of almost unsexing them.

When other faculties are active, the function is repressed through dominance of absorbing occupation, hard work, under feeding, some forms of sickness, etc. On the other hand, lux- urious idleness, absence of higher motives, general emotional indulgence and low associates degrade what is a legitimate source of happiness into its derivative, precisely as we see in the insane person all his vague feelings, his unrest from the relatively overfed tissues of the body, with absence of brain- control, tending to increase of lowest desires. The bulimia, or ravenous appetite of the insane, is the evidence of the withdrawal of the higher desires and the inordinate development of the fundamental.

Again, in others, the effect upon the actions of the cellular sexual activity may be apparent in diseases, as some forms of hysteria, while the sufferer is absolutely guiltless of any knowledge of the source of her vague desires, unrest and emotional aberrations.

Degrees of pleasurable feeling culminate in the *smile* which had primarily the useful function of enlarging the mouth preparatory to taking in a good-sized morsel. Spencer has analyzed the joyous feeling in his "Physiology of Laughter." Incongruity is at the root of ludicrousness, but the latter does not alone excite *laughter.* Pent-up feeling must find some vent.

Wonder, on the contrary, may prevent laughter; the former makes the muscles tense while the latter relaxes them. "There is a good physiological basis," says Spencer, "for the popular notion that mirth-creating excitement facilitates digestion." The visceral shaking up he notes, points, in my way of thinking, directly to the deglutitive chuckle origin of the process. The palate tickle is the root of the physiological function, and whoever has not seen the pleased grunts of the savage as he gobbles and chuckles over his food, may see something of the origin of the laugh in the pig pen while the hogs are feeding. Things which tickle have for their origin the gratified wriggle of the œsophageal and enteric cells; when the tickle occurs epidermically, laughter is produced as the only outlet toward visceral motion possible to such moleculo-molar motion, which when it occurs internally finds vent in the original cell assimilative activity.

The savage "Ugh!" or grunt of assent, testifies to both the derivation of satisfaction expression and to the relationship of

15

stomach pleasures to laughter, for until modified by cultivation of restraints, laughter is but a succession of such grunts; the scream of laughter and occasional tears of merriment are due to the overflow of energy, or the operation of Darwin's third principle of diffusion.

The dog's method of betokening pleasure by licking your hand is equivalent to telling you that you are good to eat, and it is his substitute for a smile. His fawning and cringing is an acknowledgment of your superiority, and he licks his chops, often hypocritically, in lieu of smiling at you or asserting your "sweetness." Darwin notes that some negroes more directly express this appetite association by rubbing their bellies when pleased.

Hobbe's well known suggestion of glee having its foundation in a self-superiority feeling, and the multitude of other theories with those of Spencer and Bain's, indicate that a variety of mainly pleasurable states provoke mirth. *It appears very clear to me that as the eating motions are the lowest exhibitions of pleasure, laughter has been developed thereupon, and the multitude of mental and bodily conditions inducing this expression of pleasure and other feelings influence the expression through association.*

Mobility of features and ability to express the feelings vary greatly in different persons, and it is not always safe to judge of what others feel by their expressions, for repression is developed in races which learn the inexpediency of exhibiting their thoughts outwardly as the child does. The detrimental character, of such exhibitions determines the deceit of the cloaked, inhibited expression, and even the substitution of the sickly, insincere "society smile" or smirk.

The showing of emotion by others strikes us as "greenness" and awkwardness. Nevertheless, contempt for this is mingled with admiration for the absence of deceit in the perpetrator, an admiration founded in the pleasure we experience in knowing that we may benefit by that absence, in that person.

Outcries, at first accidentally induced through diffusion, excitement of respiratory muscles, were found to answer several purposes, and, as Darwin notes, the infant has developed extraordinary ability in this line through finding how promptly it instigated relief through others. Many forms of outcries, such as yells, shrieks, etc., are still diffusion effects through inability to make more serviceable diversion of nerve excitement. The sexual selection development of singing is well known to have influenced bird evolution, and in troubadour days, and occasionally in our own times, the serenade has the same end in view as the *Lieder ohne Worte* of the cats. The evolution of language has been a simple process of phonation compounding. This is known as the "bow-wow theory," which has its proof in all ethnological and philological investigations. Recently Mr. Cushing added other proofs from his study of the Zuñi Indians.

A remarkable cerebral fact may be cited in point. A large area of the fronto-parietal part of the brain, especially on the left side, is now known by cerebral anatomists as the "symbolic field," for the reason that the gesticulations of speech, writing and other means of communication are grouped there, and destruction of these areas bring about a loss of ability to speak, write, or otherwise gesticulate, all the way from paralysis of the members to mere inability to recall a word or letter, though the ability to speak or write the word or letter exists, and can be temporarily exercised if the thing forgotten be repeated in spoken or written words. The aphasia may be for writing,—agraphia; for spoken words,—aphasia, together or separately.

Hysterical and insane persons, and young children, rudely

vent their excitement in screams, because evolved channels for expression are denied them.

Blushing, in my opinion, was originally a low nutrient reflex of the sympathetic system. Sudden impressions of nearly every kind produced a suffusion of the cephalic parts, which developed divisions of this general reflex and enabled the animals' brains to have blood sent to particular parts. Blushing is a diffusion effect not so evident in hairy or thickly cuticled animals, and both natural and sexual selection has perpetuated it through its representing *modesty*, a form of *timidity* or fear. Such visible acknowledgment of superiority accorded the male, mostly, being very gratifying to the self-conceit.

In man such, and other, diffusion evidences as tremblings, are reprobated as weaknesses and do not excite the admiration of the other sex, hence their repression has gone on in the male line with development in the female. Ontogenetic modifications have taken place in the mixing of male and female peculiarities. This modifying tendency prevents wide differences in the sexes in other things aside from direct sexual matters.

Sighing was originally, as it now is, serviceable as an inspiratory act, and directly affords relief from the oppression of imperfect oxygenation of the pulmonary blood. From other oppressed feelings clustering around this sensation through resembling it, the sigh of relief from any and all oppressed feelings has arisen as an expression for all of them. The association of feelings of oppression is a direct one, through a slowed respiration being set up by depression of the function, whatever may be the cause.

Depression has reference, then, to a respiratory interference. Dyspnœa is painful and the sigh is an effort for relief. The "love-lorn," the sympathetic, etc., sighs are all of this character.

Weeping.—With the absence from the water which had bathed

the piscine eye, the batrachian readjusted the optic capabilities by acquiring a tolerance for living on land and having occasional swims, but the lachrymal gland was formed through the posterior cells being drawn upon by the corneal and sclerotic parts until development made the fluid-furnishing function easy. The saline tears are of the nature of sea water. With this pain of exposure to light and the dessication of the eyeball, there became associated the endeavor to relieve other forms of pain by the process of weeping.

The shutting of the eyes during screaming is shown by Darwin to be serviceable in overcoming the distension of the optic by the blood sent there in the act.

Sneezing and *Coughing,* Darwin regards as originally voluntary expulsive acts of fishes, when any objectionable material was introduced into their respiratory or pharyngeal passages. From this arose the involuntary reflex acts in us for the expulsion of any irritating matter in those parts. The built up reflexes operate alike for all irritations or stimulations of an irregular nature, such as are induced by odors, congestions, accumulations of mucus, etc., where at first only the coarser materials affected those areas. This is only one of the myriad transferences of volitional to involuntary or instinctively performed reflexes.

The expression of *Helplessness,* as the cringe of man, is, like the small dog's throwing himself on his back before the large one, a serviceable attitude, and appeals to sympathy. Darwin's entire principle of antithesis I pass into his first principle of serviceable associated habit. The antithetical emotional expression did not arise as antithesis simply. That was a coincidence. But the antithetical cringe of humility was originated as a serviceable appeal to another, the serviceability being the first consideration and the antithesis to the strut of pride being the means of performing the serviceable act.

Patience, Resignation, are serviceable as appeals to sympathy and directly as conserving the strength for the things we can do from wasting them in fruitless emotional display. Helplessness is the ultra expression of both, and in abject cases passes into despair.

Suspiciousness has in it the fear of harm to self, and the furtive glance by which it is characterized comes from the desire not to be caught watching, but nevertheless a wish to closely scrutinize. The diffusion shunt apparatus of the cerebrum and other intervertebral homologues are brought into use in this feeling. When it rises to confirmation of the doubts, then the suitable reflexes to the secondary feeling appear.

Physiognomists, such as Lavater, strove to picture *Jealousy* and *Maternal Love*, but Darwin shows that there are no such expressions. Jealousy may not impart a pleasant look, but it may produce dozens of different expressions, from suspiciousness to rage. The artist, by his judicious accessories, makes the beholder think he sees maternal love in the mother's face as she bends over her infant, but wipe out the cradle and its contents from the picture, and substitute a poodle or something good to eat, and the expression will fit the change as well. The expression of love between the sexes is a masked excitability; the rapid heart-beats, when they meet, the blushes and pleased "good to eat" smile, are all this best of sentiments afford us in expression.

Parental Love has often, by writers, been ridiculously assumed as the cause of the species perpetuation. It does not take a very astute observer long to see that the getting of children holds a very subordinate part in sexual desires; often it is very undesirable. The daily community of interest and the appeals which children make upon our sympathies, their helplessness, and the rule which causes the benefactor to like the one he benefits more than the benefited likes the benefactor, operate to create parental love.

To a large extent the mother's love is a selfish one, for the infant is part of herself, its functions are so dependent upon her own as to make this love natural. Among the wealthy who neglect their children, and among the very poor who do the same through inability to do otherwise, the hearth-stone affection is not developed as in the middle classes, who, after all, are *the* anchor classes of the world.

A favor received becomes a source of irksome sense of obligation ; one imparted awakens a feeling of satisfaction in having lain another under tribute of regard. In children, however, who take everything as their due, or whose interests are bound up with their parents, the mutual love is more likely to be enduring. Ch. Ribot, and others, show that family affection between children and parents is not due to any inherent mystery, such as novel writers fasten upon it, by citing cases where bastard offspring may be tenderly loved by the father whose confidence has been betrayed, and that adopted children often arouse true parental love.

Sympathy, as Spencer shows, arises as an altruistic feeling through our conceiving ourselves in the place of the one needing the sympathy. We must have first felt the sensation in ourselves which arouses our sympathy for it in another. It is owing to this that ladies may crush bugs and flies, and the naturalist, who studies them under the microscope and realizes their kinship in pain and pleasure to ourselves, usually refrains from unnecessary infliction of pain. A vivisectionist for this reason is more likely to be merciful than the "beastiarian" who decries him. The rich, for this reason, seldom feel for the poor. In fact, he who rides in a carriage has an involuntary contempt for him who goes afoot. The knowledge that this feeling is natural should only operate toward overcoming it. Unpleasant information of this kind usually invokes a storm of denial from the mob, they prefer to think themselves descendants of the angels, and refuse to analyze their own sentiments. The unwelcome truths should be faced and an honest endeavor be made to develop good traits we do not possess.

The artistic and scientific tendencies evolved slowly from the barbarian love of display, originating in the decorative tendency of animals, still remaining to a certain extent among human females in their regard for finery and jewelry. Desire for adulation is at the root of all progress in the arts and sciences. It is a blow to the scientist to see his work appropriated without credit, and an amusing toadyism is apparent in semi-scientific circles, appropriately mentionable here, mixed with the filching tendency alluded to.

A Chicago writer dislikes to credit any one in Arkansas with a good thought. A New York or Boston man cannot conceive of Chicago originating anything, and across the sea the general run of scientists avoid any mention of America or its workers, if possible. Darwin was a notable exception to this rule, for he was above such paltriness. Quotations from the *American Naturalist* and other United States journals abound in his writings. Then the occidental toady will foster this spirit by quoting only from the East and ignoring everything American.

The average Eastward plagiarist has not been above stealing ideas from his disdained confreres, and I mention this with all the more feeling because I have seen my own work thus appropriated until I have come to the pass of noting such things with a view to their publication in the near future.

Huxley calls attention to the truly snobbish nature of the house dog, who will bark at an illy-dressed beggar, but treats the well-dressed stranger with cordiality or indifference.

These reprehensible acts arise from the mistaken notions of what constitutes expediency. The apeing, toadying, snobbery, etc., are founded in a desire to attract attention, as one way of ministering to self-conceit. Plagiarism is stupid theft, and in all an advantage is supposed to have been gained by the animal.

The views of some of the metaphysicians may be appropriately quoted with reference to the feelings or emotions herein considered : James Mill says that "when an agreeable sensation is conceived of as future, but without one's being certain of it, this state of consciousness is called *hope ;* if one is certain of it, it is called *joy.* When a disagreeable sensation is conceived of as future, but uncertain, that state of consciousness is called *fear ;* if it is certain, it is called *sorrow.* An agreeable sensation, or the idea of that sensation joined to the cause which produces it, engenders *affection,* or love for that cause. A disagreeable sensation joined to the idea of its cause engenders *antipathy,* or hatred for that cause."

According to Spinoza, " *Love* is nothing but joy accompanied by the idea of an exterior cause. *Hate* is nothing but sadness accompanied by the idea of an exterior cause."

CHAPTER XIV.

Derived Activities, Mainly Mental.

Hesitation and doubt are either, or both, lower "automatic" or mental conditions developed in proportion as we pass from the lower forms of cell life to the best coördinated organization. In the lowest form, hesitancy and doubt are due to attractive and repulsive influences, so balanced as to have not reached a resultant which will determine a motion. Indeterminate diffuse motions exhibit the greatest degree of this, and as the effect becomes more definite, then hesitation or doubt passes to another state. Doubt is an action of the diffusion apparatus, or the inability to gauge past or present impressions to action. Hesitancy is a form of doubt and is given rise to through its presence. As Belief may pass to the extent of elaborating an instinctively acting reflex system so that it may occasion "unconscious" movements, Doubt is related to Meditation, Thought, Reason, in having no such thoroughly built up apparatus.

Volition.—The end of deliberation, thought, hesitation, or doubt. It is the coördinated volitions of the cells, the resultant of their activities, and merges into the purely reflex at one end of the scale to the relatively free will of the individual with the best brain organization. Free will is relatively such only. You cannot do what your organization debars you from doing and the component cells act through natural influences. If the adjustment and associated activities of the cells be such as to enable the wisest (most expedient) action in behalf of the organism or its aims, then we have that degree of free will development which distinguishes the best intentioned (hence wisest) man from the criminal or lunatic, whose erratic, illy-ballanced volition shows him to be controlled by an imperfectly arranged association system of means to ends. In this sense the wise man is comparatively free, and the criminal and lunatic are in subjection to a badly arranged nervous system which does not unite the cells in their best interests.

The question of responsibility enters here, and is seen to be a purely relative matter, incapable of definition or fixation. The jurists have all sorts of artificial standards of responsibility and are unable to determine, except approximately, and then only occasionally, the degree of responsibility of the individual for acts committed. In the knowledge that criminality is lunacy, in that it leads the man to do things incompatible with his immediate or remote interests (and the interests of society are his own), the expediency doctrine shows us why free-booters in one age are sane, and in another criminals, and in still another insane. The irresponsibility or responsibility of those who decide upon that of others, is thus determined as a matter of no moment. They are guided by extrinsic circumstances; and the result between judge, jury and popular opinion is a juggle any way. All such relative actors and activities being in an irresponsible flux, and accident determining the "decision" in the end, which is a pure result of caprice, always.

The remotest influence may "decide" whether the arraigned shall hang or be given a fortune. The loss of a note, the state of digestion of the judge or juror, etc. And the clearest cases, apparently warranting a favorable decision, often merit a reversed treatment to that obtained. Spencer gives the best account of volition not being absolute.

Reason and Thought, to some extent synonymous, involve hesitation and doubt, where thought is not used in the sense of reverie.

Logic, or, as it has been called, the "art of reasoning," may in all its sub-divisions be included in *Deduction.* Impressions are accepted as facts, and resemblances and differences sought, whatever the sub-divisions of the deductive process may be.

The only test of truth we possess is the one mentioned by Edgar A. Poe, *Consistency.*

Instinct depends upon definition of tracts in the nervous system, and hence succeeds reason, though inherited or other instinct may be disintegrated through tearing down of tracts by reason and subsequently reinstated as instinct again of a different kind.

Reverie, may be conceded to be an indulgence in memories with but little of a definite end in view. Reasoning implies a doubt, and preceding determination is a species of hesitation as palpable as that which precedes the choice of a reflex from the spinal cord and is of the same nature. In the latter case, the diffusion fibrillæ of the cord, or the neuroglia, or both, are acting indeterminately. In the case of reasoning and in *Meditation* the brows are often contracted, as though to shade the eyes, and respiration slowed as in attention, the feeling is allied to pain and associated with anticipation and the desire to concentrate the perceptions, often apparent in occasional elevation of the eye-brows and other attention attitudes. The diffusion fibers of the brain are allowed full sway, and according to the revivability of past experiences in terms of memory sensations the result of the deliberation will be evident. Through incessant reaction to certain stimuli a spinal cord reflex may be inevitable. This depends upon a definite arrangement of the reflex adjustments to the stimulus and constitutes instinct. If the stimulus be novel and adjustment be not fully made to it, and more than one mode of automatism be possible, then the hesitation in the choice of reflexes is the reasoning of the cord. If no provision for action is practicable, then diffusion is the only result and the tremblings and other general diffusion manifestations occur. Similarly with the brain workings. If a certain line of action is the invariable result of a certain precedent impression, then the higher instinctive motions or reflexes, whether simple or complex matters not, result. If two or more acts or modes of expression are possible as a result of the deliberation, then the cerebral diffusion apparatus will "choose" one or the other exhibition, the choice being a predetermined resultant of a set of conditions the most likely to be associated under existing circumstances of blood supply, brain structure and previous modifying influences.

If no provision for determination is possible, then the diffusion apparatus finds no outlet in activity for its stimulation and diffusion. Effects similar to those furnished by the cord follow, and owing to the greater blood consumption and the preponderance of brain tissue over spinal cord tissue, so will the results of painful indecision be proportionably greater than those attending the lower centers. In extreme cases, exhaustion, mental anguish, psychic pain, which is generally the same as bodily pain and relieved by the same medicaments, or frenzy, may be the result.

Wonder is expressed by the facial contortions of attentive perception. In proportion as the wonder is greatest, the diffusion apparatus in the vicinity of the cerebral gyri involved in the perception are suffused with blood through the nutrient reflex action. As soon as memory or recollection drops the thing wondered at to a cognized or familiar matter, wonder ceases, and a resultant motion of the body is made, or the former activities are resumed. In this, and in perception, according to their depth, the nutrient reflex assists the associating methods, which, when effected, other actions succeed.

Wonderment has been, and is, powerful in constructing the diffusion fibrils of the brain, which in turn cohere in fasciculi to form the radiating tracts, and in the human infant and simian the wonder faculty distinguishes them as reflecting animals, which revolve in their minds the possibilities of what they perceive being convertible to their own uses. The infant who thrusts everything into its mouth, shows what its ideas of utility are, and the chimpanzee is divided, in his inspection of the novel thing between doubts as to its injuriousness or edibility, as is the infant later in life.

In fact, the majority of men, whether "civilized" or savage, lose all interest in matters which do not refer to a pretty low sense gratification, and it is only by development of the reasoning (cerebral fibræ arcuatæ) abilities that attention can be sustained over things which involve higher considerations—such, for instance, as abstract good to a people, instead of concrete good to one's self.

The *Imagination* is an exercise of memory, the superimposition of recollections, the recalling of images of things and acts, and the vividness of these images depend upon the exercise of the faculty and the reasoning powers owned by the individual. It is but a form of thought, and is definite or indefinite in what it yields, according to the cerebral structure of the person. It has Reverie at one extreme of its indulgence and profound *abstraction* at the other.

Ideas are resultants of thought operations. A fixed idea or conviction is a belief or conception of an actuality judged as such by the person's action conforming thereto. James Mill recognized ideas as mental reproductions of sensations.

The *desire for Liberty or Freedom* is not an exalted feeling, for it is identical with the locomotory desire of the amœba, which we said was associated inextricably with hunger in that animal and separated from it only when locomotory apparatus was differentiated. It, however, has been evolved from the concrete selfish to the abstract altruistic love of national freedom.

The *Æsthetic feelings* Spencer has shown to be dependent upon the "*play desire*" of lower animals. The wish to expend "nervous" energy in rythmic or other motions. Dancing and the pleasure derived from music belong to this feeling, the universality of rythm in nature, and particularly in our own motions and sensations inspire the love of music and timed motions, the drama, etc.

Belief, Confidence, Trust, Assurance, Conviction, concededly relate to action. "Preparedness to act upon what we affirm is admitted on all hands to be the sole, the genuine, the unmistakable criterion of belief."* It is based upon an assumed knowledge, which in turn depends upon memory. Bain also says, "it never occurs to the child to question any statement made to it until some positive force on the side of scepticism has been developed."

Prescience depends upon memory, and is corroborated in precise proportion as the knowledge of past facts or impressions justify the prediction of their recurrences and the acuteness of the individual in seeing and foreseeing relationships.

Forethought falls in this faculty of prescience.

My treatment of the sensory and molar activities in this essay is open to the charge of a want of profounder classification. I have purposely withheld the usual divisions of these phenomena, principally because such separations of unconscious and conscious acts, feelings, desires, emotions, and mental states, are purely artificial; and next, because it would require too much space to treat such matters fairly. In preference to presenting abstract groupings which would have

---

* Bain, op. cit., p. 505.

included all the activities in a general way, the plan adopted seemed more conducive to a clearer understanding. A description of the evolution of a few of the states in the order of their development has been attempted, and I have been forced to reserve further consideration of matters relating to the special psychology of man for a bulkier volume, to follow the publication of this.

CHAPTER XV.

THE LAW OF EXPEDIENCY AND OPTIMISTIC CONCLUSION.

Bain.* gravely mentions Benjamin Franklin's "Moral Algebra, or method of deciding matters for one's self," by setting down all the *pros* and *cons* and striking a balance, the preponderance of the remainder deciding the step to be taken. *Prudential.* algebra is another name for the process.

With the operation of the law of expediency under our eyes every minute, and its active shaping of all animate affairs, neither Bain nor his predecessors have been able to see its application in Franklin's method, nor in the multitude of similar processes resorted to unconsciously, but naturally, every day by every animal.

What is expedient to do must be determined by every animal, according to his circumstances and his way of thinking. His ideas of what may be desirable will guide him usually, and whether one desires food, another its representative, money, or another a perpetual continuance of possession of both in immortality, the expediency gauge of the individual may be applied by considering *what* he deems desirable and the practicability of obtaining it. Considerations of the proper methods for the accomplishment of an end also involve questions of feasibility and expediency. For it would not do to destroy an end in the method of its attaining. Yet, through bad logic or the want of mastery of the emotions or desires, frequently this is done.

---

* The Emotions and The Will, p. 413.

16

The advantages arising from a continuance in "doing right" become so apparent as to make that action often habitual, whereupon the animal takes great credit for his *conscientious-ness.*

A friend of mine, whose learning and the value of whose opinions are known the world over, was formerly beset by lawyers and judges in expert cases. They considered it expedient to annoy him by scurrility, as it amused the court room mob and reacted against him with the average juryman. Finally, the judges and the attorneys were reminded that while they temporarily raised a laugh, such a thing as printer's ink existed. The ability of my friend to defend himself against a recurrence of such insults, acted as a deterrent upon their exuberance, and finally induced them to consider *common decency* expedient. Every animal has had to learn the lesson, and the highest are willing to profit by the experience of others, and not undergo the usual preliminary discomfitures.

Conscience has arisen from this law of expediency, and no matter whether the thing to be dreaded as a consequence of the act has a real existence or not, so long as there are unpleasant consequences, or the animal imagines there are, flowing from certain acts, the disagreeable association of the acts with their real or imagined outcome constitute that feeling of depression or anticipation of evil accompanying the individual standard of bad action. The pleasure derivable from a good act is similarly associated with a reward, such as all ecclesiasts have found it expedient to hold out to all men as a consequence of a certain line of action. By inheritance, the habit of decency may become second nature to a person, and the advantages are so evident to him that unless circumstances conspire to destroy the feeling, he prefers to "do right," because it is easier to do so.

The penny given to the beggar whom you may never see again, awakens your self-approbation. It is your reward. If you have inherited benevolence you feel " guilty " if, when the opportunity arises, you omit doing a kind act. Your desire to escape the disagreeable consequences of having throttled your kindly promptings may retrace your steps and cause you to make a sacrifice as a " sop to Cerberus."

Most charitable acts are perfunctorily performed, and the public conscience is assuaged through its alms by proxy. The tax-payer, from whom the stint has been wrenched by law, usually refers the alms seeker to the commissioners for the poor, and satisfies his " conscience " in a way which could not be done were the public to be depended on to do its own alms-giving.

Only a few animals and men have correct conceptions of the highest expediency. The majority act up to the crudest, shallowest ideas. Thus, anger exhibition and stupid selfishness characterize their dealings, where the few control themselves and have exalted aims in life.

Many who see what is expedient are not able to govern their passions and are placed at the same disadvantage as those who are blind to their best interests.

The history of the animated world consists of the individual and collective triumph of higher over lower expediency. It may be seen in the wiles and doublings of the chase; the survival of the fittest in multitudes of instances; the gradual elevation of the standard of fitness as expediency adopts a higher plane; selfish individual savage scrambles and brawls are superceded by tribal and national defensive and offensive coherence against enemies; the interweaving of interests all over the world through the extension of commerce and rapid communication put an end to international quarrels, and with the growth of wisdom (expediency recognition) happiness is increased and will eventually reach a plane which, though relative and incomplete, will be incomparably greater for all than can now be conceived.

Herbert Spencer's early writings were pervaded by attacks against the expediency philosophy, but later he tacitly adopted it. His original idea was the omnipotency of a "moral law," and he made a sincere attempt to expound the Christian ideal of such a law. Many of his notes on this topic are touching and beautiful. For instance, he points to the fact that the moral law assigns no punishment for its infringement; it directs us to what should be done by way of elevating mankind, but is silent concerning penalties. This seems very like a recognition of universal irresponsibility, and is quite reconcilable with both adjurations to forgiveness and the chemical affinity theory of life which pervades this book.

If a cell or organ may, under certain favorable circumstances, be developed into any other kind of cell or organ, and as we positively know to be the case, circumstances largely control the character of men, it is folly indeed to be otherwise than forgiving, seeing that we are none of us responsible in any particular. But that this cannot at present be adopted generally comes from the unevolved condition of human nature and society. A flexible arbitrary standard of responsibility will exist as long as true expediency does not govern all. Penalties and rewards are accompaniments of all Nature's laws. Pains will be minimized and the pleasures increased as fast as natural laws are recognized and allowed to govern action, until finally the highest adjustment at present conceivable will evolve a prevailing gentleness, honesty, justice, forgiveness, and an altruism which will prove to be an exalted egoism because the best interests to all will be subserved through its governance.

I claim that *Expediency is the Moral Law.* Several hundred years ago "stand and deliver" was a legitimatized salutation, and the booty obtained no more afflicted the average conscience than the looting of a public treasury to-day would interfere with the ordinary politician's digestion. Mercantile interests gradually extended protection and the free-booter gave way to trade knaveries, such as adulterations, embezzlements, etc. A good old German novel, "*Soll und Haben,*" idealizes commercial integrity in the character of a merchant who kept an account with God into which he passed all his debits and credits, and who tried to make his dealings consistent with the supposition that the deity would punish all unfairness with his fellow men.

Such characters will prevail numerically and in their successes, eventually, as a higher ideal of expediency is created. Spencer shows that when money-getting ceases to be the chief interest of life, and nobler aims are substituted, the world will look back with horror at the conceptions of happiness in this age, and "honesty in tatters" will be regarded as a martyr among the "successful" thieves of this generation, while in generations to come honesty will not be allowed to be the unfittest to survive. Natural penalties will swiftly follow the derelictions of low-grade expediency worshippers and the rewards will be more justly distributed.

Teach children to be upright because it is wisest so to be, and show them how it befalls that respect, position and honor accrue from ordinary decency, and such instruction is likely to have more weight than all the usual indefinite promises and appeals to their fears that can be made.

The saying that "every man has his price" means more than its vulgar interpretation. If you seek broad and worthy fields of life and can accustom yourself to derive happiness from noble work (I will not attempt to define what this may be, for though ideals may differ, a generally fair understanding of what nobleness is, exists); if retrograding to a state of parasitism or developing an overweening taste for wealth, which is founded upon the gustatory and has its associations therewith so closely that an unscrupulous capitalist is merely a hypertrophied intestine; if becoming cruel and selfish are repugnant to your ideals of being happy, then your better longings, in proportion to their intensity and the consistency with which you endeavor to satisfy them, places you beyond the bartering of your higher expediency pleasures for that which the social sharks, foxes and tape-worms scramble.

Nevertheless, the robber and one who works for "eternal rewards" differ only in degree of expediency ideals.

The reason why good eventually triumphs is simply because development of intellect renders people capable of seeing what is truly expedient, that same faculty is associated with the ability to thwart the low grade intellects of those whose expediency conceptions are lower. Besides, in innumerable ways, evil tends to destroy itself, even though it may require thousands of years and involve the destruction of many unfit to exist nations.

" But," the affrighted mind cries out, "what consolation is all this to me as an individual? The world may in its endless cycles be growing better, and my childrens' children may, in living up to the true expediency, realize happiness, but what is that to *me ?* I want some assurance of the future!"

This is the natural lower feeling, the genuinely selfish atomic affinity for the most neighboring kindred atoms. The contemplation of death is pleasant to no one. It may be, even seems likely, that, with the evolution of the conceptions of true happiness and the better acquaintanceship with the inflexibility of Nature's laws, more resignation to them will be acquired, even to the extent Bulwer describes in his "Coming Race."

The question has been asked: "Does the amœba ever die?" When we regard the countless multiplications of the initial form and consider that the offspring are parts of the parent, the living in one's children is the only kind of immortality we can observe. More than this it is folly to affirm in the present state of knowledge; yet, as Huxley says, the audacity of those who claim to know all about the hereafter is only equaled by the arrogance of those who deny its existence. Agnosticism, which is the wise refraining from questions which we cannot settle, is our only refuge.

We may take comfort in the feeling that, as that which is universally good succeeds inevitably over that which is bad, the operation of such a law will not in the end overlook the individual. Only let us strive to deserve a higher kind of happiness.

INDEX.

www.ingramcontent.com/pod-product-compliance
Lightning Source LLC
Chambersburg PA
CBHW020354030726
47496CB00007B/2127